T0210770

SpringerBriefs in Energy

Energy Analysis

Series Editor

Charles A.S. Hall, Professor Emeritus, SUNY College of Environmental Science and Forestry, Syracuse, NY, USA

The "energy crises" of the 1970s together with the appearance of numerous books and papers on the general theme of "limits to growth" catapulted energy from obscurity to social, economic, and academic prominence. Then fuel prices came down again, economies recovered, and energy more or less disappeared from university and political discourse until about 2005. Many believed that market processes had resolved, and would continue to address, all energy supply issues. Now energy in all of its aspects is back with a vengeance. While oil spills, mine deaths, and stock market plunges grab the headlines, there are much more fundamental discussions taking place about energy and its economic and environmental effects, many already occurring, on a societal scale. Most fundamental in this respect is the arrival, plus or minus a few years, of global peak oil, which in 1968 was first predicted to occur around the turn of the millennium. Clearly, energy never really went away and in fact underlies all physical motion, all life, all chemistry, and all economics . It has been neglected in our understanding and teaching of societal processes for far too long, and particularly now when the scope and importance of issues growing out of energy availability and use are increasing yearly. These issues and impacts include the potential for economic growth and wealth creation (including so-called sustainable development), climate change, general pollution, agricultural production, clean water, and perhaps even the continued existence of civilization as we know it in the coming decades.

We believe that there is a great need for a comprehensive and integrated series of books that provides a quantitative accounting of energy use including the potential and limitations of real-world deployments of new technologies. One key concept is Energy Returned on Energy Invested (EROEI or EROI) which is used to quantity the actual net energy for society from sources as diverse as "renewable" energy from solar and wind power installations to the current enthusiasm over unconventional oil and gas. **Springer Briefs in Energy Analysis** will cover the fundamental ways in which energy operates in the natural world and, in its abundance and governing physical laws, enables and constrains all human activities. The series will be empirically based with much technical information while remaining accessible to the non-specialist reader. Individual books will minimize the jargon, mathematics, or theory of specialist books in the various disciplines on the one hand, and the advocacy positions of popular accounts that may make their scholarship and conclusions suspect on the other.

Under the editorship of Charles Hall and a panel of energy experts, the series will include contributions from many of the world's most authoritative energy analysts. The series will appeal to anyone interested in energy use and its impacts on the economy and society in general.

The audience will range from advanced undergraduates through professionals in the physical science, environmental science, and economics and financial communities. Financial analysts will gain an understanding of how energy is increasingly impacting economic processes. University instructors will find these books to be invaluable for providing students (and themselves) with greater depth and insight into the role of energy in society with an emphasis on the methods and applications of energy accounting.

More information about this subseries at https://link.springer.com/bookseries/10041

Patrick Moriarty · Damon Honnery

Switching Off

Meeting Our Energy Needs in A Constrained Future

Springer

Patrick Moriarty
Department of Design
Caulfield Campus, Monash University
Caulfield East, VIC, Australia

Damon Honnery
Department of Mechanical and Aerospace
Engineering
Monash University
Clayton, VIC, Australia

ISSN 2191-5520   ISSN 2191-5539 (electronic)
SpringerBriefs in Energy
ISSN 2191-7876   ISSN 2199-9147 (electronic)
Energy Analysis
ISBN 978-981-19-0766-1   ISBN 978-981-19-0767-8 (eBook)
https://doi.org/10.1007/978-981-19-0767-8

This Springer imprint is published by the registered company Springer Nature Singapore Pte Ltd.
The registered company address is: 152 Beach Road, #21-01/04 Gateway East, Singapore 189721, Singapore

Contents

Abbreviations

AFOLU	Agriculture, forestry and other land uses
BECCS	Bioenergy with carbon capture and storage
CBA	Cost-benefit analysis
CC	Climate change
CCC	Catastrophic climate change
CCS	Carbon capture and storage
CDR	Carbon dioxide removal
CFC	Chlorofluorocarbon
DAC	Direct air capture
EGS	Enhanced geothermal systems
EIA	Energy Information Administration (US)
EJ	Exajoule (10^{18} J)
EROI	Energy return on investment
ESME	Ecosystem maintenance energy
EW	Enhanced weathering
FF	Fossil fuel
GHG	Greenhouse gas
GNI	Gross national income
Gt	Gigatonne (10^9 t)
GW	Gigawatt (10^9 W)
HANPP	Human appropriation of net primary production
IAEA	International Atomic Energy Agency
IAM	Integrated assessment model
ICEV	Internal combustion engine vehicle
IEA	International Energy Agency
IPCC	Intergovernmental Panel on Climate Change
ITER	International Thermonuclear Experimental Reactor
Mt	Megatonne (10^6 t)
NET	Negative emission technology
NPP	Net primary production
OECD	Organisation for Economic Co-operation and Development

OPEC	Organization of the Petroleum Exporting Countries
OTEC	Ocean thermal energy conversion
p-k	Passenger-km
PPP	Purchase parity pricing
RCP	Representative Concentration Pathway
RE	Renewable energy
SCC	Social cost of carbon
SDG	Sustainable Development Goal
SMR	Small modular reactor
SRM	Solar radiation management
STEC	Solar thermal energy conversion
SUV	Sports Utility Vehicle
UHI	Urban heat island

Chapter 1
Introduction to Global Energy Challenges

Abstract Intentional energy use by humans is many millennia old, but until about two centuries ago, the energy used was almost entirely bioenergy. Since then, we have witnessed a transition to fossil fuels as the dominant energy source. This change has produced not only many benefits but also a number of environmental side effects, chiefly climate change from emitted carbon dioxide. In this introductory chapter, we outline the serious challenges we face in attaining ecological sustainability, their interconnections, and their links to the global energy future. On a national per capita basis, both energy use and consequent carbon emissions are very unevenly distributed, largely the result of inequitable income distribution. Paradoxically, those countries with the lowest carbon emissions per capita are those most likely to bear the brunt of the effects of climate change. Besides climate change, other problems that have indirect impacts on energy, and need to be urgently addressed, are declining biodiversity, global chemical pollution, and continued global population growth.

Keywords Climate change · Climate emergency · Energy costs · Energy history · Energy inequality · Fossil fuels (FFs) · Global environmental problems · Global pandemic · Renewable energy (RE)

1.1 Introduction: A Brief History of Energy

Renewable energy (RE) is not a new idea; humans first used the combustion of wood around 500 millennia ago [11], for warmth, light, cooking food, and later for smelting metals and firing bricks. (Not that combusted biomass was the first use of energy, as all animals have an existential need for food energy. Deliberate energy use for food—by both grazing animals and other animals that preyed on them—preceded human use by hundreds of millions of years. Further, we may not even be the only species using fire. In northern Australia, it is reported that various species of hawk drop burning embers from a bushfire ahead of the front, to start new blazes to drive out small prey [3].) Nor was wood fuel the only RE source humans used, as apart from the muscle power of humans and their domesticated animals, wind energy was used for powering watercraft, and later, both wind and water mills were built for providing mechanical energy. In a sense, even geothermal energy was used, as humans have enjoyed the use of geothermal pools for thousands of years [28].

Fossil fuel (FF) use is likewise not new. 'Some forms of petroleum, coal, and natural gas were used thousands of years ago by various civilizations on various continents, according to historical records and archaeological finds' [19]. Natural gas was used in China two millennia ago, and bamboo pipes were used to carry the gas to buildings for both heating and lighting. Britain, the home of the industrial revolution, was using some coal in Roman times [19]. Nevertheless, FF use was tiny compared to the universal use of biofuels. In 1800, as the Age of Fossil Fuels got underway, global coal production had risen to around 10 million tonnes, or about 5% of all primary energy.

Figure 1.1 shows how spectacular the growth in global energy use has been since around 1900, when FF and RE sources had equal shares. The growth in FFs in different regions has been far from uniform, as FF use has actually declined in a number of Organisation for Economic Co-operation and Development (OECD) countries, but has risen manyfold in China over the past three decades [4]. It's clear that in recent years, the upturn in RE use has been paralleled by rising use in FFs; far from substituting for FF, at a global level at least, the two energy groups appear to complement each other. Further, most RE is still derived from fuel wood, burnt at low efficiency in low-income countries, especially in tropical Africa; modern biomass use is still small. As a result, commercial RE in 2020 was still only 12.6% of all commercial energy—but is up from the 11.4% figure in 2019, according to BP data [4].

The IEA [16] forecast a 4.6% rise in global energy use in 2021, giving a 0.5% growth over 2019 levels. (Global GDP also contracted in 2020, but an expected 6% rebound will lead to a 2% rise over 2019 pre-Covid levels.) One piece of good news concerns RE: unlike FFs, RE output grew during 2020 by 3% and seems set to rise

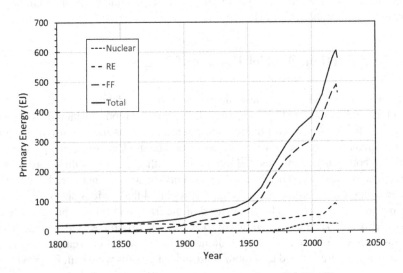

Fig. 1.1 Energy use from 1800 to 2020 by energy category. *Source* [4, 27, 28]

again in 2021. The bad news is that coal demand is also expected to grow strongly in 2021. Over 80% of the 2020 FF decline is expected to be reversed in 2021, so that FF CO_2 emissions will be just 1.2% below those for 2019 [16]. The FF comeback implies that commercial RE will decline slightly from its 2020 share of total primary energy.

1.2 The Perilous State of the Planet

As has been well-documented, the spectacular rise in global FF use has been accompanied by a rise in anthropogenic greenhouse gas (GHG) emissions to the atmospheric commons, with carbon dioxide (CO_2) by far the most important of the energy-related GHGs, although methane (CH_4) emissions are also significant. When combusted, around 69.1 megatonnes (Mt) of CO_2 are presently emitted for every exajoule (EJ = 10^{18} J) of the current global fossil fuel mix [4]. Increasing use of lower CO_2-emitting fossil fuels such as natural gas (CH4) in the energy mix will see this amount fall, but methane leakage and fugitive emissions must be suppressed. According to the Elsevier Scopus database in June 2021, over 9200 refereed papers had both of the terms 'fossil fuels' and 'climate change' contained in either the title, abstract, or keywords, while for 'climate change' alone, the figure was nearly 322,000.

Researchers are increasingly using the terms 'catastrophic climate change' (CCC) and 'climate emergency' to describe both the seriousness of the CC problem and the urgent need for effective mitigation actions [30]. The first Intergovernmental Panel on Climate Change (IPCC) report was released in 1990. In the three decades since, the reports have become progressively more urgent on the need for action. The sixth report bluntly states: 'It is unequivocal that human influence has warmed the atmosphere, ocean and land. Widespread and rapid changes in the atmosphere, ocean, cryosphere and biosphere have occurred.' In 1990, the average annual atmospheric CO_2 concentration measured at Mauna Loa, Hawaii, was about 353 parts per million (ppm), but by August 2021 the monthly average was 414.5 ppm [31]. Evidently, neither the numerous international climate conferences nor the 322,000 papers mentioned earlier have made any significant impact on policies to curb CO_2 emissions. The lack of action on climate change (CC) has not resulted from a lack of research. Alas, the climate seems to respond more to the concentration of the various greenhouse gases (GHGs) in the atmosphere, rather than the number of papers and books (including our own) urging GHG reductions.

One criticism of the IPCC Fifth Assessment Report was that it did not adequately acknowledge the existence of tipping points in the climate system, an omission corrected in the latest report [14]. Folke et al. [10] identified a set of 15 climate-related *tipping points*. Lenton et al. [23] examined nine of these, most of them referring to polar or other high-latitude regions, such as ice loss from the Greenland icecap. They reported that the tipping points were not independent of each other: if one was crossed, it made others more likely to be crossed. One consequence they foresaw was that Earth might be irreversibly committed to 10 m or so of rising seas, even if the rise occurred slowly over centuries. Plainly, this is a burden we will place mainly on future generations. The authors also stressed that the time left before catastrophic changes was shrinking, especially compared to the time taken to develop such an effective policy response. As the sixth IPCC report puts it: 'Life on Earth can recover from a drastic climate shift by evolving into new species and creating new ecosystems. Humans cannot.'

According to the IPCC [14], the temperature has risen linearly with atmospheric CO_2 ppm in recent decades, but ecological changes expected are not linear with temperature. The IPCC warn that the ecosystems will change in a non-linear and possibly unpredictable manner with further rises in temperature. The adverse ecological effects observed as the rise in global surface temperature since pre-industrial times increased from 0.5 to 1.0 °C were much greater than those observed with the earlier rise from 0 to 0.5 °C. Even more serious changes to ecosystems such as coral reefs can be expected from the next 0.5 °C temperature rise, already underway. The 2021 temperature of 1.1 °C above the 1850–1900 average is the highest since the interglacial 125 millennia ago, driven by atmospheric CO_2 levels at their highest for two million years. The 1.1 °C rise is a global average: land surface temperatures have risen more than the oceans, and the Arctic region has experienced increases far above average. The rapid changes occurring to the Arctic are a foretaste of what more temperate (and populous) regions will experience—unless we take drastic action.

On land, extreme heatwaves that once occurred every half-century can probably now be expected every 3 or 4 years should global temperatures rise above 2 °C. We can also expect to experience two or more extreme events occurring simultaneously, such as coupled droughts and heatwaves. Ominously, forests, soils, and the oceans will most likely in the future be less able to mop up emitted CO_2, meaning more will remain in the atmosphere.

Bad as it is, climate change isn't the only environmental problem we need to worry about. Rockström, Steffen, and colleagues [34, 37] have developed the concept of planetary *boundaries*; breaching any of these limits can endanger global sustainability. The original 2009 list of the limits was as follows:

1. climate change,
2. rate of biodiversity loss,
3. nitrogen cycle and phosphorus cycle change,
4. stratospheric ozone depletion,
5. ocean acidification,
6. global freshwater depletion,
7. change in land use,
8. atmospheric aerosol loading, and
9. chemical pollution.

For only one of these limits, stratospheric ozone depletion, was there any movement away from the limit. In a later paper [37], the list was slightly modified, and further attempts were made to define numerical limits for each of the nine boundaries. For climate change, the safe limit for atmospheric CO_2 was set in the range of 350–450 ppm. If the present growth rate continues, we are likely to pass the 450 ppm upper limit in little more than a decade. Even the present value may be too high for safety, as the spate of extreme floods, heatwaves, and wildfires in 2021 attest. Biodiversity loss is also regarded as being at a critical stage, as its proposed limit has been transgressed [10], as we explain below.

Just as we saw for the various climate system tipping points, multiple interactions are possible between the various planetary boundaries [2, 13]. Heinze et al. [13] warn that there is a high probability of crossing several ocean tipping points like ocean warming, acidification, and deoxygenation, all with serious consequences for marine ecosystems. These multiple interactions will act to shrink the 'safe operating space' for any individual boundary. As an example of such feedback, consider the declining yields predicted for important food crops under progressive global warming. One result could be increased deforestation for new agricultural land, which would not only raise CO_2 emissions but also lead to further biodiversity loss [20].

Bradshaw et al. [5] have also argued that—apart from the risk of catastrophic CC—our planet not only faces a number of other challenges, but that these have been greatly underestimated, even by many scientists. The most important of these in their view is the global decline in biodiversity. For vertebrate species with population sizes that have been monitored over the past half-century, an average decline in populations of 68% has been observed. Recorded vertebrate extinctions—which probably underestimate the real count—over the past 500 years have averaged around four every three years, which is at least 15 times the background extinction rate. The planet has already seen five major extinctions since the Cambrian period. The authors assert that we are now on the path leading to a sixth major extinction—and the first one involving humanity as both cause and possible casualty.

Why is biodiversity collapse important? Because, according to Sage [35] it affects 'not just species loss, but a loss of the ecological interactions, functions, redundancy, co-dependencies, structural complexity, and mechanisms of resilience that characterize living systems and their ability to provide ecosystem services'. We do not even know how many species currently inhabit Earth; one estimate among many is 8.7 million eukaryotic species, the great majority of which are yet to be studied [26]. Other scientists caution that the biodiversity picture is more nuanced: most species show either very small changes in numbers, with the overall decline due to a few outlier species. Nevertheless, they still agree that biodiversity loss, especially change at the local level, is serious. The fear is that the rapid rate of species replacement at the local level will destabilise ecosystems [39].

The biodiversity and the CC challenges to sustainability are interdependent [10], and so, too, are their solutions. Without a 1.5 °C limit on temperature rise, biodiversity losses will be too great; conversely, if the downward trend in biodiversity is not reversed, temperature rises will be too high. One obvious connection between biodiversity loss and climate change is through the ongoing decline in global forested areas, with consequent loss of both species and carbon stored in trees. Further, greater biodiversity in tropical forests enhances their carbon storage [40]. According to Folke et al. [10]: 'The diversity of functional groups and traits of species and populations are essential for ecosystem integrity and the generation of ecosystem services.' If there is a variety of species performing similar functions, the ecosystem will be more resilient to shocks or extreme events, such as changes in drought frequency and intensity resulting from climate change. Like CC, biodiversity loss and ecosystem loss of resilience have been effectively ignored.

At the beginning of the current era, humans numbered perhaps 150 million, and even by the start of the twentieth century, our numbers had only risen to around 1560 million [22]. The human population in 2021 is nearing 7800 million and continues to rise [10]. As an example of human dominance, one estimate is that 10 millennia ago, humans—together with their livestock and companion animals—only comprised a tiny fraction of the combined mass of all mammals; today the value is 96% [10]. Because of resource pressure and inequitable distribution of global resources, food insecurity is also rising. As many as 700–800 million people receive insufficient daily calorie intake, and perhaps one to two billion lack the micronutrients necessary for proper functioning. The current pandemic is only making it worse for many people already living on the edge. With the global population set to rise further to around nine billion by 2050, we could be facing the prospect of widespread famine [5], unless we make major changes to agriculture, diet, and food distribution ethics. The increasing number of people with chronic food insecurity are also a threat to biodiversity loss: they may have little alternative but to clear forested land, and try to feed themselves regardless of ecological consequences.

One threat which many observers think should be placed alongside CC and biodiversity loss is chemical pollution, the last entry in the global limits listed above. Thousands of synthetic chemicals have been added to the environment. Many of these are harmless, many are not; we just don't know in many other cases [21]. The UN Environmental Program, in their 2021 report *Making Peace with Nature* [38], summarised the human cost of pollution: 'In 2015, it was estimated that 6.5 million deaths were attributable to ambient and indoor air pollution combined, 1.8 million deaths to water pollution (unsafe water sources and inadequate sanitation) and another 1.3 million deaths to pollution stemming from soil, heavy metals, chemicals and occupational environments.' This chemical pollution threat is evidently not just some future challenge. Ironically, the air pollution from fossil fuel combustion was largely responsible for the hiatus in global warming until the 1970s. Burning FFs, especially coal and oil, produces various sulphur oxides, as well as CO_2. The reduced insolation from pollution offsets the temperature rise from CO_2 increase. Air pollution from FF use is often still severe in industrialising countries, but as it is cut for health reasons, additional net climate forcing will occur.

One solid pollutant that is the cause of rising concern is plastics, which are mainly made from natural gas or petroleum. Their presence in the environment adversely affects the health of both humans and natural ecosystems through organism 'entanglement and ingestion'. Plastic pollution is growing worldwide, with marine plastics pollution 10 times higher than in the year 1980 [38].

Beginning in early 2020, the Covid-19 pandemic has dominated both news and government policy. The pandemic provides a perfect illustration of how developments in one sector—in this case public health—can have a huge impact on all others— including the energy sector. It has significantly reduced energy use, most notably aviation turbine fuel use, as air travel was cut back. Biodiversity loss is also connected with global public health: the global biodiversity decline has increased our risk from new zoonotic diseases like Covid-19 [17, 18]. Or, as McCallum [25] has tersely put it in his eponymous article: 'Lose diversity, gain disease'.

1.3 Everybody, Everywhere Needs Energy

Although international data shows that while average GDP per capita for various countries is more equal than a few decades ago, the same is not true for income (or wealth) at the individual level. Alvaredo et al. [1] have examined such changes between 1980 and 2016. They found that the richest 1% adults in the world accounted for 27% of total global income growth over the period. In stark contrast, the bottom half of all adults only gained 12% of this total. The result was that the top 1% income earners increased their share of the total from 16 to 20% between the years 1980 and 2016.

Even as little as a century ago, per capita primary energy use in different parts of the world did not differ much, although in the more developed countries, particularly in Western Europe, fossil fuels accounted for a large share of the total. National incomes also varied little on a per capita basis. A few decades earlier, in 1870, the estimate for Sub-Saharan Africa overall was US$ 800 (in 2020 values) compared with a then world average of US$ 1500 [24]. The lower income countries relied almost entirely on biomass, which was still usually plentiful there on a per capita basis. Today, although per capita primary energy use (and many other indicators of human welfare [29]) vary greatly from country to country, energy inequality is far more starkly revealed by electricity use per capita, where the data is also much more reliable, and mirrors income inequality. For a start, Word Bank data [47] showed that in 2019, 10% of the global population, or nearly 800 million people, had no access to electricity— 'switching off' was not an option for them! Transport energy inequality is illustrated by the fact that in many countries, most people have never been on a plane. Elon Musk is the entrepreneur CEO of Tesla, Inc., which makes the Tesla car, an environmentally friendly electric vehicle (EV). In just one landmark year, in 2018, Musk flew 150,000 miles (241,000 km), equivalent to circling the Earth six times at the equator, in his petroleum-fuelled private jet [12], which nicely illustrates the other extreme in energy use.

Table 1.1 shows the average per capita national electricity consumption in selected countries, including those with the highest and lowest use, for the year 2018. Per capita electricity consumption in the leading country, Iceland, is over *1000 times* larger than for Haiti, the lowest for which data is available—and data-poor countries are nearly always very low energy users. Not surprisingly, those countries with low electricity use also had both low GNI/capita and low CO_2 emissions per capita from energy and industry.

For many pollutants, reductions can be achieved by actions at the local or regional level, as with urban air or freshwater pollution. This is not the case for anthropogenic global warming. Even if a middle-sized high-income country, such as Germany, were

Table 1.1 Electricity, energy CO_2 emissions, and economic data for various countries, 2018-19

Country	[1]GNI/capita	[2]kWh/capita	[2]Tonne CO_2/cap
China	16,760	4906	6.84
Eritrea	[3]1610	90	0.15
Ethiopia	2310	83	0.12
Haiti	3040	37	0.29
Iceland	59,590	54,605	6.22
India	6920	968	1.71
Japan	43,880	8010	8.55
Norway	70,810	24,047	6.78
US	66,080	13,098	15.03
World	17,541	3260	4.42

Source Ref. [15]
[1]PPP = Purchase parity pricing (US$ 2019)
[2]2018 data
[3]2011 value

to cut its CO_2 emissions to zero, it would do little to stem local climate change in Germany—or anywhere else; CO_2 is a well-mixed gas in the atmosphere. Ironically, those countries that have historically contributed least to the problem of CC will suffer disproportionately from its consequences [36]. The reason? Most of the low-energy use, low-income countries, lie in the tropics (see, for example, Eritrea, Ethiopia, and Haiti in Table 1.1), regions where plants, animals, and humans are already much closer to their thermal tolerance limits than is the case for temperate climates. A further problem, as Perez and Stroud [32] point out, is that tropical species, whether plant or animal, have in general evolved to have a narrow thermal tolerance range, because of the low climate variability that characterises much of the tropics. Even in high-income countries such as the USA, low-income (which are usually also low CO_2-emitting) neighbourhoods suffer disproportionately during heatwaves [46].

Table 1.2 shows, for China and the US (the two leading energy use nations in 2020), and for the world overall, commercial energy use (i.e. excluding fuel wood) and CO_2 emissions from FFs and industry for years 1965 and 2020 [4]. For both primary energy and CO_2 emissions, China has risen from a minor position in 1965 to the top position in both by 2020. The US, the leading country for both in 1965, now lies in second place, and, like a number of other mature industrial economies in the OECD,

Table 1.2 Energy use (EJ) and CO_2 emissions (Gt) for China, US, and the world, 1965 and 2020

Country/region	1965 EJ (%)	2020 EJ (%)	1965 CO_2 (Gt) (%)	2020 CO_2 (Gt) (%)
China	5.5 (3.5)	145.5 (26.1)	0.49 (4.4)	9.90 (30.7)
US	52.4 (33.7)	87.8 (15.8)	3.48 (31.1)	4.46 (13.8)
World	155.7 (100)	556.6 (100)	11.21 (100)	32.28 (100)

Source Ref. [4]

both its energy use and emissions are declining [4]. (Much of the observed OECD decline is not the result of energy conservation or efficiency improvements, but rather results from de-industrialisation and consequent imports of energy-intensive manufactured products from non-OECD countries like China.) Nevertheless, the OECD countries have historically been responsible for most cumulative energy-related GHG emissions, globally estimated at nearly 2400 Gt over the period 1850–2019 [14]; in 1965, their share was still almost 70% of the global annual total [4]. But today even the enlarged OECD only accounts for around one-third of energy-related emissions, and its share is declining. (When net deforestation is also factored in, their share of total anthropogenic CO_2 emissions is even smaller, since OECD forest carbon storage is rising.) Were the OECD countries to cease all GHG emissions tomorrow, Earth would still be heading for climate catastrophe. Deep emission reductions are needed in most countries, not only the OECD.

The 'carbon pie' is defined as the CO_2 we can safely emit before causing serious anthropogenic CC. According to the latest IPCC report [14], if we are to have two chances in three of staying within the 1.5 °C limit, the 'carbon pie' remaining after the start of 2020 was only 420 Gt of CO_2. The year 2019 energy-related CO_2 emissions were 34 Gt [4], or little more than a decade's worth. Various approaches are possible for a fairer distribution of this ever-dwindling quantity. Because CO_2 gas has a long half-life in the atmosphere [14], cumulative emissions over the past century or so are relevant for equity. At the national level, the distribution could be based on the Egalitarian Principle, with equal per capita emissions for all countries by, say, the year 2050. 'The two secondary allocation principles, termed Emission-based and GDP-based, distribute allowable emissions according to cumulative historical emissions and cumulative historical GDP respectively' [7]. However, neither of the latter two principles would give equal average per capita emissions by 2050.

All three approaches are based on *national* averages: an even more radical approach could address the vast inequality in carbon emissions between *individuals*, regardless of where they live. As Folke et al. [10] reported: 'the wealthiest 1% of the world's population have been responsible for more than twice as much carbon pollution as the poorest half of humanity.' Such an approach would require knowledge of individual emissions. A similar approach has actually been considered for the UK: The Personal Carbon Trading scheme. 'This scheme proposed to allocate equal tradeable carbon credits to all eligible individuals (who may simply be taken here as all adults), with the aim of reducing carbon emissions, as part of an emissions trading scheme' [44]. The scheme would also address income inequality at the national level, but could also be extended internationally, which would give low-energy use consumers, mainly in low-income countries, surplus carbon credits that they could sell [29].

However, one possibility that must be considered is that ongoing global warming could make such national allocations increasingly less relevant in a decade or two. Most GHG emissions today are either from FF combustion or deforestation. While the resulting GHG emissions can clearly be assigned to a given country, what about CO_2 and methane losses from melting permafrost in boreal regions? Or, CO_2 emissions from forests because of wildfires, drought, and pests? Or, methane emitted from ocean sediment clathrates? Even if many fires are the result of human actions, many are from lightning strikes, and such losses will be exacerbated by a worsening climate. Although GHG releases from all these sources are presently minor, their potential releases from a variety of feedbacks, including the impact of increased forest fires on permafrost releases [6], could eventually be comparable to, or even dwarf anthropogenic emissions.

1.4 Discussion and Conclusions

The Earth is in a thermal imbalance, with insolation exceeding back-radiation into space. Earth is heating up, and one consequence we are already experiencing is a rise in extreme weather events, as predicted by climate models. Henan Province in China experienced flooding in July 2021; the event was described as a once in 1000 year flood, with the city of Zhengzhou experiencing over 200 mm of rain in a single hour [45]. In mid-2021 an unprecedented heatwave hit N-W America, with the Canadian national temperature record broken by an astonishing 5 °C. Scientists calculated that the chances of this heatwave were made 150 times more likely by climate change. The once in a 1000 year heatwave took climate modellers by surprise. As one wryly commented at the time: 'We are much less certain about how the climate affects heatwaves than we were two weeks ago' [42]. In the American west, the frequency of droughts will rise, in turn increasing the intensity and frequency of wildfires [43]. The rise in extreme weather—particularly floods and droughts—will also negatively impact global food production [9].

Already, Raymond et al. [33] have documented: 'The emergence of heat and humidity too severe for human tolerance'. They argued that some places on Earth already have wet-bulb temperatures exceeding 35 °C, the upper physiological limit for humans. The areas unfit for human habitation will expand as global temperatures continue to rise. Attempts to maintain habitability by extensive use of air conditioning will increase energy use and consequently GHGs, and will therefore exacerbate global warming [41]. But air conditioning only goes so far; we are already seeing evidence of 'climate refugees' as people migrate from increasingly uninhabitable regions.

Large cities, which are home to a large and still rising share of the human population, face a further thermal challenge. The Urban Heat Island (UHI) effect results from several causes, including heat release from all energy-using devices such as vehicles and air conditioners, reduced vegetative evapotranspiration cooling from sealed surfaces such as roofs and pavements, and reduced radiative heat loss to space because of the 'canyon effect' of tall buildings. The result: UHI can make cities warmer than the surrounding countryside—sometimes by several °C [28]. UHI will be a benefit in winter for temperate climates, but will exacerbate summer heatwave frequency and intensities.

Climate change is only the most obvious—and discussed—challenges to our existing energy system. As shown, a variety of other threats to ecological stability exist. One difficulty is that the other global limits do not carry warning signals such as the extreme weather events experienced by more and more of the world's population. Nor do they have an easily measured (and understood) indicator like atmospheric CO_2 concentration. Some of these other threats will directly impact energy—the damaging impacts of drought on biomass or hydro energy production are examples. Others will have indirect effects, possibly by exacerbating climate change. As an example of interaction, some observers feel that the present pandemic is an early sign of the consequences of biodiversity loss. When all the threats to biosphere integrity are considered together, the conclusion must be that we are entering unchartered territory. As indicated on old maps: Here be Dragons!

The remaining four chapters of this book cover the following topics. Given the global problems outlined in this chapter, Chap. 2 examines the prospects for technological solutions, including boosting nuclear power output, energy efficiency improvements, and two largely untried technologies, carbon dioxide removal and geoengineering by solar radiation management. These solutions are mainly designed as responses to climate change. The prospects for greatly increased uptake of the various forms of RE are treated in Chap. 3. Chapter 4 builds on the earlier chapters to discuss the possible global futures for energy. Finally, Chap. 5 goes beyond technological solutions to the challenges facing energy production and use by looking at more fundamental questions. How much energy do we really need? What for? What policies need to be introduced if we are to build a just and ecologically sustainable planet? And finally, what are the chances for success in this endeavour?

References

1. Alvaredo F, Chancel L, Piketty T et al (2018) The elephant curve of global inequality and growth. AEA Pap Proc 108:103–108
2. Barnosky AD, Brown JH, Daily GC et al (2014) Introducing the scientific consensus on maintaining humanity's life support systems in the 21st century: information for policy makers. Anthropocene Rev 1(1):78–109
3. Bonta M, Gosford R, Eussen D, et al (2017) Intentional fire-spreading by "firehawk" raptors in Northern Australia. J Ethnobiol 37 (4): 700–718. https://doi.org/10.2993/0278-0771-374700
4. BP (2021) BP statistical review of world energy 2021: 70th ed. London, BP
5. Bradshaw CJA, Ehrlich PR, Beattie A, et al (2021) Underestimating the challenges of avoiding a ghastly future. Front Conserv Sci 1: 615419. https://doi.org/10.3389/fcosc.2020.615419
6. Chen Y, Romps DM, Seeley JT et al (2021) Future increases in Arctic lightning and fire risk for permafrost carbon. Nature Clim Change 11:404–410
7. Ekanayake P, Moriarty P, Honnery D (2015) Equity and energy in global solutions to climate change. Energy Sustain Dev 26:72–78
8. Fischer EM, Sippel S, Knutti R (2021) Increasing probability of record-shattering climate extremes. Nature Clim Change 11:689–695
9. Fitzgerald T (2016) The impact of climate change on agricultural crops. In: Batley J (ed) Plant Genomics and Climate Change; Edwards D. Springer, Berlin/Heidelberg, Germany, pp 1–13
10. Folke C, Polasky S, Rockström J, et al (2021) Our future in the Anthropocene biosphere. Ambio 50: 834–869 https://doi.org/10.1007/s13280-021-01544-8
11. Gowlett JAJ (2016) The discovery of fire by humans: A long and convoluted process. Phil Trans Roy Soc B 371(1695) https://doi.org/10.1098/rstb.2015.0164

12. Harwell D (2019) Elon-musk's highflying 2018: What 150000-miles in a private jet reveal about his excruciating year. Washington Post: https://www.washingtonpost.com/business/economy/elon-musks-highflying-2018-what-150000-miles-in-a-private-jet-reveal-about-his-excruciating-year/2019/01/29/83b5604e-20ee-11e9-8b59-0a28f2191131_story.html
13. Heinze C, Blenckner T, Martins H, et al (2021) The quiet crossing of ocean tipping points. PNAS 118 (9): e2008478118.
14. Intergovernmental Panel on Climate Change (IPCC) (2021) Climate change 2021: The physical science basis. AR6, WG1. CUP, Cambridge UK (Also earlier reports)
15. International Energy Agency (IEA) (2020) Key world energy statistics 2020. IEA/OECD, Paris
16. International Energy Agency (IEA) (2021) Global Energy Review 2021. https://www.iea.org/reports/global-energy-review-2021
17. Keesing F, Ostfeld RS (2021) Impacts of biodiversity and biodiversity loss on zoonotic diseases. PNAS 118 (17) e2023540118
18. Khetan AK (2020) COVID-19: Why declining biodiversity puts us at greater risk for emerging infectious diseases, and what we can do. J Gen Intern Med 35(9):2746–2747. https://doi.org/10.1007/s11606-020-05977-x
19. Kool T (2020) The complete history of fossil fuels. https://oilprice.com/Energy/Energy-General/The-Complete-History-Of-Fossil-Fuels.html
20. Lade SJ, Steffen W, de Vries W et al (2020) Human impacts on planetary boundaries amplified by Earth system interactions. Nat Sustain 3:119–128
21. Lawton G (2021) Earth's chemical crisis. New Sci 24: 36–42
22. Lehman C, Loberg S, Wilson M, et al (2021) Ecology of the Anthropocene signals hope for consciously managing the planetary ecosystem. PNAS 118 (28): e2024150118
23. Lenton TM, Rockström J, Gaffney O et al (2019) Climate tipping points —too risky to bet against. Nature 575:592–595
24. Maddison Project Database 2020 (2020) https://www.rug.nl/ggdc/historicaldevelopment/maddison/releases/maddison-project-database-20200
25. McCallum HI (2015) Lose biodiversity, gain disease. PNAS 112(28):8523–8524
26. Mora C, Tittensor DP, Adl S, et al (2011) How many species are there on Earth and in the ocean? PLoS Biol 9 (8): e1001127 https://doi.org/10.1371/journal.pbio.1001127
27. Moriarty P, Honnery D (2011) Rise and fall of the carbon civilisation. Springer, London
28. Moriarty P, Honnery D (2019) Energy accounting for a renewable energy future. Energies 12:4280
29. Moriarty P, Honnery D (2020) New approaches for ecological and social sustainability in a post-pandemic world. World 1:191–204. https://doi.org/10.3390/world1030014
30. Moriarty P, Honnery D (2021) The risk of catastrophic climate change: energy implications. Futures 128: 102728
31. National Oceanic and Atmospheric Administration (NOAA) (2021) Trends in atmospheric carbon dioxide. https://www.esrl.noaa.gov/gmd/ccgg/trends/
32. Perez TM, Stroud JT (2016) Thermal trouble in the tropics. Science 351:1392–1393
33. Raymond C, Matthews T, Horton RM (2020) The emergence of heat and humidity too severe for human tolerance. Sci Adv 6: 18–38
34. Rockström J, Steffen W, Noone K et al (2009) A safe operating space for humanity. Nature 461:472–475
35. Sage RF (2020) Global change biology: A primer. Glob Change Biol 26:3–30
36. Schiermeier Q (2016) Telltale warming likely to hit poorer countries first. Nature 556:415–416
37. Steffen W, Richardson K, Rockström J et al (2015) Planetary boundaries: Guiding human development on a changing planet. Science 347(6223):1259855
38. United Nations Environment Programme (2021) Making Peace with nature: A scientific blueprint to tackle the climate, biodiversity and pollution emergencies Nairobi. https://www.unep.org/resources/making-peace-nature
39. Vaidyanathan G (2021) The world's species are playing musical chairs. Nature 596:22–25

40. Van der Sande MT, Poorter L, Balvanera P et al (2017) The integration of empirical, remote sensing and modelling approaches enhances insight in the role of biodiversity in climate change mitigation by tropical forests. Curr Opin in Environ Sustain 26–27:69–76. https://doi.org/10.1016/j.cosust.2017.01.016

41. Vaughan A (2020) Some places are already too hot for humans to live. New Sci 247:19

42. Vaughan A (2021) The heat is on out west. New Sci 10 July: 10–11

43. Vaughan A (2021) Climate change made heatwave more likely. New Sci 17 July: 11

44. Wang SJ, Moriarty P, Yi Ming Ji YM et al (2015) A new approach for reducing urban transport energy. Energy Proc 75:2910–2915

45. Wikipedia (2021) Henan floods. https://en.wikipedia.org/wiki/2021_Henan_floods

46. Witze A (2021) The deadly impact of urban heat. Nature 595:349–351

47. World Bank (2021) Access to electricity (% of population). https://data.worldbank.org/indicator/EG.ELC.ACCS.ZS

Chapter 2
The Problems with Tech Fixes

Abstract Technical fixes are highly favoured by most decision-makers because they involve the least disruption to existing social and economic order. However, in today's 'full world' they often meet with unintended consequences. In this chapter, we examine a number of these in the context of mitigating climate change: nuclear power; energy-efficient improvement; various technologies for carbon dioxide removal (CDR); and geoengineering in the form of solar radiation management (SRM). Nuclear energy is losing market share, and even the nuclear industry does not predict share recovery. Reductions in energy intensity have not prevented global energy growth, because of unmet demand in presently low-energy countries. CDR in the form of forestation has been implemented in some places, but net loss in global forest biomass is still occurring. Other forms of CDR are still unproven at the very large scales needed. SRM is likewise unproven, and like CDR technologies, would eventually face FF depletion.

Keywords Afforestation · Carbon dioxide removal (CDR) · Climate change · Direct air capture (DAC) · Energy efficiency · Nuclear power · Solar radiation management (SRM)

2.1 Introduction: Unintended Consequences

In an essay entitled *The Great Horse-Manure Crisis of 1894* [15], Steven Davies wrote: 'In New York in 1900, the population of 100,000 horses produced 2.5 million pounds of horse manure per day, which all had to be swept up and disposed of.' The problem was similar in London, where 'one writer estimated that in 50 years every street in London would be buried under nine feet of manure'. As we now know, such dire forecasts did not come to pass, and Davies subtitled his essay *'The problem solved itself'*.

True, the end of the horse transport era brought many benefits to the US. Not only from the disappearance of horse manure on the streets, with their accompanying clouds of flies, and smell, but also from the freeing up of the vast area of land needed to grow hay for the thousands of horses, as well as their stabling areas in cities. However, again with hindsight, it seems we have merely substituted one set of

P. Moriarty and D. Honnery, *Switching Off*, Energy Analysis,
https://doi.org/10.1007/978-981-19-0767-8_2

problems with another. The vast rise in the internal combustion engine vehicle (ICEV) fleet produced air and noise pollution, and, given its near-total reliance on petroleum fuels, fears about energy security. Even more recently, as discussed in Chap. 1, the risk of climate change, especially from FF combustion, has led to a search for non-carbon vehicle fuels, to solve the new problem for a technical fix which solved the earlier problem. Paradoxically, one of the alternative fuels promoted and widely used today was biofuels. However, biofuel in the form of hay was the fuel *already* used by horses. (Energy analyst Cesare Marchetti [33] also claims that the energy consumption in the form of hay by the 30 million working horses in the US in 1920 would roughly match the energy used if cars had replaced the horses.)

The horse manure saga is only one example of tech fixes that solved one set of problems only to create a fresh set. Chlorofluorocarbons (CFCs) replaced ammonia and sulphur dioxide as a refrigerant in the late 1920s, because it was considered much safer, given its low reactivity, toxicity, and boiling point [37]. Later, in 1985, it was discovered that CFCs cause depletion of the ozone layer, which protects humans and other species from ultraviolet radiation. They have now been largely replaced by more ozone-friendly chemicals. CFCs are also extremely potent GHGs—as are their replacements [24]. As another example, *tetraethyl lead* was introduced into gasoline in order to improve the fuel's octane rating and engine compression, enabling an economical way of raising fuel efficiency. Much later, it was found that lead pollution from vehicles was a serious health problem, especially for residents living alongside heavily trafficked city roads. (As an interesting aside, both the use of CFCs as a refrigerant and leaded petrol were developed by the same person, chemical engineer Thomas Midgley Jr [65], who tragically died by strangulation after being entangled in a hoist system he invented to lift him out of his bed after contracting a debilitating illness.)

Edward Tenner, in his entertaining 1996 book *Why Things Bite Back: Technology and the Revenge Effect* [57], provides numerous other examples of solutions that backfired. But the realisation of unintended consequences of introducing new technologies or policies is far from new; it was extensively treated by Stanley Jevons in his 1865 book *The Coal Question*, in which he first drew attention to what is now known as the rebound effect in energy demand. It may well be the case that whatever success tech fixes had in the past, they now ever-increasingly come attached with significant drawbacks. Table 2.1 gives a list of several innovations to help the environment, which ended up making things worse.

A possible reason for this can be found in the rise of the global limits, discussed in Chap. 1. The planet is simultaneously approaching a variety of planetary limits—not just those needed for climate stability. A solution to one problem is increasingly likely to move the world closer to one or other of the limits. Or, as noted environmental economist, Herman Daly [14], has stressed, we now live in a 'full world', not the 'empty world' of even a century or so ago, with a world economic output product only a fraction of that today. Solutions possible in the year 1900, a world of only 1.56 billion and around 296 ppm atmospheric CO_2 levels, are not possible in the crowded world of 2021, with 7.8 billion people and 414 ppm CO_2 levels [42].

Table 2.1 Examples of environmental innovations: intended and unintended consequences

Innovation	Intended consequence	Unintended consequence	References
Cane toads—Australia	Destroy cane pests	Poisoning of native wildlife	[46]
CFC refrigerants	Safe refrigerants	Stratospheric ozone destruction	[57]
DDT pesticide	Kill crop pests	Poisoning of non-target species	[57]
Gypsy moth caterpillars	Silk industry in the US	Widespread deforestation	[46]
Leaded petrol	Improve octane rating	Lead poisoning in urban areas	[57]
Sparrow cull—China	Reduce rice crop damage	Insect pests increased crop damage	[67]
Computers	'Paperless office'	Office paper use increased	[57]
Antibiotics	Reduce infection	Rise of antibiotic resistance	[57]
Football helmets	Reduce risk of injury	Behaviour change-rising injuries	[57]
Tall smokestacks	Prevent local pollution	Increased distant pollution	[67]

It is easy to see why tech fixes are the preferred solution to our many environmental and resource problems. First, we have the stunning developments in Information Technology (IT) as an inspirational example. Computer guru Ray Kurzweil [30] even introduced the idea of 'The Singularity', to be brought about by the exponential advances in technology. As he put it: 'This is a time when the pace of technological change will be so rapid and its impact so deep that human life will be irreversibly transformed. We will be able to reprogram our biology, and ultimately transcend it. The result will be an intimate merger between ourselves and the technology we are creating.' In this view, technology can solve all problems. A second important reason is that technical fixes fit in well with a business-as-usual growth economy, as they require minimal changes to existing social or economic systems.

In this chapter, we look at several proposed solutions to the climate change problem thrown up by our heavy use of fossil fuels. These include greatly increased use of nuclear energy, profound improvements in energy efficiency, carbon dioxide removal (CDR), both in its biological and mechanical variants, and geoengineering in the form of solar radiation management (SRM), all treated in turn in the following sections. Another important—and heavily promoted—solution, a shift to non-carbon sources of energy, is treated in Chap. 3.

2.2   Nuclear Power

Nuclear power was born out of the 1940s successful Manhattan project to build nuclear fission weapons. After atomic bombs were dropped on Hiroshima and Nagasaki in 1945, civilian nuclear power was seen as the peaceful use of the atom. The first commercial nuclear plants date back to the 1960s, and by the mid-1970s forecasts on nuclear prospects were highly optimistic, as evidenced by an International Atomic Energy Agency (IAEA) global forecast made in 1976 of over 2500 GW in the year 2000 [29]. Since nuclear power in the US was seen—in the celebrated phrase—as 'too cheap to meter', such forecasts were not surprising. The actual global capacity in the year 2000 was only a modest 348 GW.

In 2020, nuclear energy has a 10.1% share of global electricity generation, down from its peak of 17.5% in 1996. In absolute terms, the global 2020 output of 2700 TWh was still slightly lower than the 2006 value. Most nuclear energy is still generated in OECD countries, but since around 2012, OECD output has stagnated. In contrast, nuclear output in China, which had no nuclear output before 1993, has risen rapidly, and has now passed French output, moving to the second place [8]. Table 2.2 gives the five leading countries in 2020, together with their output in 1990. Japan ranked third in 1990, but the Fukushima accident lead to a curtailment in nuclear output.

What caused the stagnation in global output? In the OECD countries, which in the 1970s and 1980s produced the bulk of nuclear power output, it was caused by a combination of factors. First was the citizen opposition, especially in Europe and the US, regarding reactor safety, waste disposal, and nuclear proliferation. This opposition was heightened by the nuclear accidents at Three Mile Island in 1979 in the US, Chernobyl in 1986 in the then Soviet Union, and at Fukushima in Japan in 2011. These were merely the most serious—and most publicised—accidents: a full list shows over 50 less serious events in the US alone [66]. Figure 2.1 shows a generalised nuclear fuel cycle for fission reactors.

In recent times, an important factor affecting nuclear power prospects has been market liberalisation, which forced nuclear plants to compete economically with other energy sources, especially gas-fired turbines and renewable energy. Although nuclear power is a low-carbon source, in the absence of significant carbon taxes, its high capital costs and long lead times for construction lower its attraction for

Table 2.2 Annual nuclear power output (TWh) in selected countries, 1990 and 2020

Country	1990 TWh	2020 TWh
US	607.2	831.5
China	0.0	366.2
France	314.1	353.8
Russia	118.3	215.9
South Korea	52.9	160.2
World	**2000.5**	**2700.1**

Source [8]

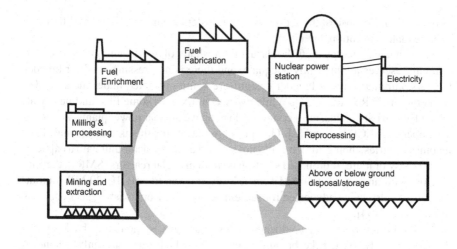

Fig. 2.1 Schematic diagram of the nuclear fuel cycle. Reproduced from Fig. 6.2 in [36]

private utilities. A related factor is cost reductions for wind and especially solar electricity. Nevertheless, all energy sources receive subsidies to some extent [38], making economic comparisons difficult. Even today, all reactors under construction in the world are being built with some level of direct or indirect government financial support.

Nuclear advocates can point to two undeniable advantages of nuclear power: it is a low-carbon emission source, and it produces dispatchable electricity (unlike intermittent wind and solar energy), although nuclear reactors are better run as base load plants. Early on, it was realised that if nuclear power was to be a major and long-lasting energy source, it would be necessary to use *breeder* reactors. Unlike conventional (or thermal) fission reactors, breeder reactors generate more fissile material than is consumed. Fission reactions produce a neutron flux, as well as radioactive by-products. Fast neutrons are needed for breeding, so that water cannot be used as a coolant (as in the popular light water reactors), since it 'moderates' (slows down) the neutrons. Fast neutrons enable the 'fertile' U-238 isotope to be transmuted into the 'fissile' plutonium-239 (Pu-239) isotope.

The advantage of breeder reactors becomes evident when it is realised that natural uranium consists of only 0.7% of the fissile isotope U-235. The rest is mainly fertile but non-fissile U-238. So, breeder reactors differ from conventional thermal reactors (such as the light water reactor) in two important ways. First, the spent fuel from the reactor must be extracted from the reactor core, cooled, then reprocessed to chemically extract the fissile Pu-239 for reuse as a reactor fuel. This extraction not only is expensive but also opens up the possibility of a country diverting plutonium for nuclear weapons production under cover of a civilian nuclear program. Second, since water can no longer be used as a coolant, other coolants which do not moderate the neutron flux from fission reactions are needed, with liquid metals such as sodium as one option. Because of these differences, breeders are inherently more difficult to operate than conventional reactors.

Since nuclear energy now holds a minor share of world energy production, uranium reserves are presently adequate, and there is no pressing need for breeder reactors, although several breeder designs are being promoted under the so-called 'Generation IV Reactors' [34]. While every country uses some FFs, and nearly all have FF electricity plants, most countries do not have nuclear power plants, including a number of OECD nations [8]. One possible reason for this is that in small low-income countries, their grid size could not support a large reactor unit with a capacity of 1000 MW or more. A proposed solution is small modular reactors (SMRs), thermal reactors with an output of 300 MW or less, even down to only 10–20 MW output. Their use is also being advocated for settlements remote from existing grids, such as mining towns [34].

SMRs are not a new idea—dozens of SMR designs, some dating back several decades, already exist. Like breeder reactors, of which there are only a handful still operating, SMRs have not taken off commercially. One reason is economies of scale. In general, output capacity for industrial plants (in this case TWh/year) increases faster than costs; usually, larger reactors produce cheaper electricity [49]. For nuclear reactors in general, Ramana also stresses that, unlike other energy sources (for example, photovoltaic solar energy), the learning curve appears *negative* for nuclear energy in general. The more we learn about reactors, the more expensive they are to build.

A radically different approach to nuclear power is a *fusion* reactor. In fusion reactors, light atoms are fused together—in this case the hydrogen isotopes deuterium (D) and tritium (T). As for both breeder reactors and SMRs, the approach is not novel: Post [48], writing a half-century ago, optimistically foresaw commercial fusion energy by the mid-1980s. In the early 2020s, assessments are much more subdued, with major commercial use not forecast until the second half of this century [34]. A coalition of leading nuclear energy nations is funding a demonstration fusion plant at Cadarache in France; this plant (the ITER plant) is intended to show the technological feasibility of fusion power. Claessens [12] reports in his book that cost estimates have risen at least four-fold, and estimates a cost in 2020 values of around 40 billion Euros. The date for delivering D-T reactions, merely the first step for commercial feasibility, has likewise been pushed back to around 2035.

Given these construction time and cost overruns, a number of fusion researchers are starting to question whether the tokamak approach to fusion—as used at Cadarache—is the most suitable. Hassanein and Sizyuk [21] titled their article '*Potential design problems for ITER fusion device*'. Bradshaw et al. [9] had earlier pointed to another problem: the issue of whether the necessary materials would be available in sufficient quantity. Different, cheaper, experimental fusion devices are being built, but whether they could exhibit commercial feasibility is in question. In summary, large-scale commercial fusion power is probably more than 50 years away, if ever, and the reactors will be very difficult to operate. In summary, fusion energy, at least on Earth, cannot be a short- or even medium-term solution for our energy problems;

in the meanwhile, we'll have to make do with the energy produced via the fusion reactions taking place on Earth's Sun.

Despite being a low-carbon energy source, nuclear energy has one important similarity to FFs; both mortgage the future [34]. For FFs, this occurs because once combusted, the CO_2 emitted, as a long-lived gas, will affect climate for many decades to come. Similarly, U-238 and U-235—the main constituents of natural uranium ore—have half-lives measured in billions of years, and so are not very radioactive, just as fossil fuels left in the ground have few environmental effects. After fission in a reactor, the radioactivity of the fission products is high, with half-lives typically measured in tens to hundreds of years. (The intense radioactivity of very short half-products requires holding spent fuel rods in cooling ponds for some time before they can even be considered for permanent burial.)

Even the organisation set up to promote nuclear power, the International Atomic Energy Agency (IAEA), does not expect nuclear energy to have more than a minor role in the year 2050 [26]. IAEA projects a range of 5.7–11.2% for the nuclear share of global electricity output in 2050; the upper estimate is a little above the 10.1% present share. The costs, timing, and environmental benefits of nuclear power have probably been too optimistic, so that nuclear energy is unlikely to have more than a small share of future global energy.

2.3 Improving Energy Efficiency

Nuclear energy is one low-carbon energy source, and Chap. 3 will examine the various forms of renewable energy, the fastest growing energy source. Here, we look at a very different approach—cutting energy use by greatly improving energy efficiency. Since all energy sources have some environmental costs, energy efficiency makes sense, especially since it is usually the cheapest approach to cutting GHGs and other emissions. Further, several researchers claim that energy efficiency can play a major part, in not only tackling energy-related environmental problems but also any concerns about energy security. They stress the vast difference between the theoretical efficiency possible for energy-consuming devices and present efficiency levels. Noted energy efficiency expert Amory Lovins has coined the word '*negawatts*' to make the point that energy savings are a cheap alternative to building new energy capacity [41].

It is one thing to point to efficiency potential, but another to bring about large *absolute* energy cuts through this approach. We have earlier discussed the notion of energy rebound as an unintended consequence of efficiency gains. For road passenger travel, improvements in vehicle energy efficiency appear to lead to both improved vehicle performance and larger vehicles. One explanation is the rising popularity of Sports Utility Vehicles (SUVs). In the US in 2019, SUVs accounted for nearly 72% of all new light-vehicle sales [16]. SUVs are also increasing their market share in the more densely populated countries of Europe and Asia. The end result is that transport energy efficiency gains are not occurring anywhere near fast enough. According to data gathered from 16 IEA countries, including the major economies, surface passenger transport intensity (MJ/p-k) only fell 10% over the period 2000–2018, not enough to counter the rise in global travel from increased global vehicle ownership [25, 44].

This move to more energy-intensive forms of transport is general, and is partly the result of slower modes being progressively replaced by faster modes. As Gabrielli and von Kármán pointed out many decades ago, vehicle speed is gained at the expense of energy efficiency. Motorised public transport replaced walking, cars replaced public transport, and air travel replaced car or rail for long-distance passenger travel, with each step incurring an energy efficiency penalty. It is nevertheless true that the energy efficiency of each specific type of transport has improved—but not enough to overcome the transport mode upgrades [35].

A major obstacle for efficiency-led energy cuts is the vastly uneven per capita incomes in different countries, discussed in Chap. 1. The consequence is deep differences in ownership of all energy-intensive goods, such as private vehicles and domestic white goods, and in air travel. The average world cars/1000 population is only about one-quarter of that for the high-income OECD countries [44]. Apart from cars, the real costs of domestic white goods such as refrigerators and air conditioners are declining, enabling purchase by more and more people in low-ownership countries [39]. There is thus a huge unmet demand for all these goods and services, which is likely to swamp any specific efficiency gains. Global average per capita energy use increased by 17% over the period 2000–2019 [8]. In Vietnam, for example, where there has been a surge in manufacturing, and consequently per capita wealth, over the same period per capita energy use increased almost five-fold.

Other factors blunt the energy reduction potential of more efficient devices in what is a global growth economy. Energy efficiency is only one of several efficiency measures. Others include capital efficiency, time efficiency, and land use efficiency. Further, as Fouquet [18] has shown, real energy prices are at historically low levels, whereas everyone is still limited to 24 h per day, and the Earth's surface is also not expanding. So, time efficiency (for example, measured by travel speed or hourly production rate) and land efficiency (for crops, measured as yield in tonnes per hectare) have risen in relative importance. Also, industries that are heavy users of energy (e.g. aluminium smelting) have already taken steps to improve energy efficiency, leaving little scope for further improvement. And for technology that depend on thermodynamic cycles, such as electricity production from fossil fuels, physics ultimately limits what is possible. As these limits are approached, gains made come with increasing system complexity and so cost.

Another important factor to consider for household energy is the progressive mechanisation of household and garden tasks, already well advanced in the OECD countries. Domestic appliances have replaced manual methods, and similarly in the garden, with lawnmowers and hedge clippers, and that annoying innovation, leaf blowers. But a further push for rising energy use comes from the continual development of entirely new products, a necessity in a growth economy. Think of the home and office computers and printers, including their standby energy use, the high energy use of server farms, and the high energy costs associated with bitcoin mining

[32]. We have also seen the rise of bottled water for drinking replacing tap water. In this case, water delivered by pipeline, an efficient form of transport, is replaced by energy-intensive small-truck deliveries.

A very different consideration arises for the case of transport energy efficiency, and possibly to some extent in other energy sectors. Even if vehicles used no energy at all for propulsion, they would still cause environmental damage. Very little of the Earth's remaining wilderness areas lie far from a road. As the first author has previously written [35]: 'These roads cause millions of animal and bird fatalities each year, barriers to wildlife movement, and edge effects for both flora and fauna species.' Highway and vehicle lights also cause night-time light pollution. The use of salt for de-icing roads in cold climates finishes up in streams and rivers, where it can cause damage to fish. The de-icing salt also causes damage to road structures, such as steel bridges. In these cases, it is not the transport energy use itself, but the energy-using devices and their necessary infrastructure which damage the environment.

In conclusion, we must continue to improve the energy efficiency of all our devices, but, on their own, efficiency improvements are unlikely to help reduce the environmental burden of energy use. Our business culture places efficiency in high esteem, but only to the extent that it helps profits. Bottled drinking water may be energy inefficient, but it has made a branded commodity out of a generic product that formerly cost next to nothing and produced a vast quantity of waste from the 1 million water bottles that are currently opened every minute and which largely remain unrecycled once used. (Fig. 2.2 shows that global growth in per capita plastic production has outstripped growth in per capita global energy supply over the period 1970–2015, leading to the possibility of this age being called the '*Plastocene*' rather than

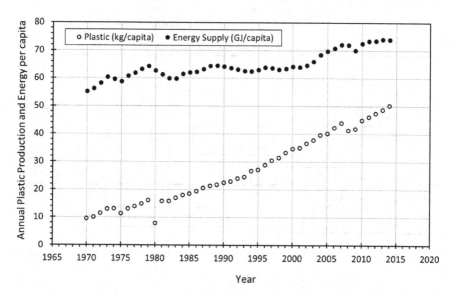

Fig. 2.2 Global annual per capita plastic production and energy supply during 1970–2015. *Source* [8, 50]

the *Anthropocene.*) Greatly raising the monetary costs of energy as a driver towards greater energy efficiency would have a big impact on use, but would at present be inequitable, since energy subsidies, both hidden and overt, assist low-income households the world over (see Chap. 5).

2.4 Carbon Dioxide Removal (CDR)

A very different approach from either using less FF energy, or through efficiency improvements, or by replacing FFs with low-carbon sources such as nuclear energy is to remove CO_2 from the atmosphere. The approaches are of two main types: biological and mechanical. The biological approaches include increased carbon storage in agriculture, forestry and other land use (AFOLU) sectors, such as by afforestation, reforestation, and various innovations to increase soil carbon. The key task, of course, is to halt the present ongoing progressive deforestation, especially in the tropics. The second approach uses mechanical means to either capture CO_2 from the exhaust gases of power stations and other large sources of emissions such as oil refineries, or to capture CO_2 directly from the atmosphere. Figure 2.3 shows the various options for CDR in schematic form.

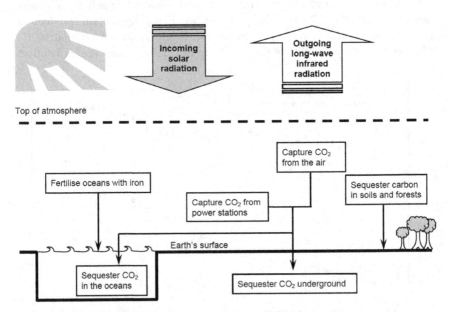

Fig. 2.3 Schematic diagram of carbon capture and sequestration options. Modified from Fig. 8.1 in [36]

2.4.1 Biological Sequestration

Several researchers believe that biological CDR can be a major solution for addressing climate change [e.g. 4, 20], mainly through reforestation. Given the heavy loss of both forest and soil carbon over recent centuries, increased carbon storage could also be seen as partially restoring Earth to its earlier position. In contrast, other researchers are sceptical that such practices can be of more than minor benefit [e.g. 40, 47, 63], or argue that they carry risks [52]. Veldman et al. [63], in their critique of the very optimistic Bastin et al. [4] paper, claimed that this paper has overestimated the global sequestration potential by as much as a factor of five. An important reason was their inclusion of savannas and grasslands for afforestation. However, not only are such areas either already productively used by humans, or important for biodiversity, but the existing soils and vegetation already sequester large amounts of carbon.

Particularly in high-latitude regions, the increased tree cover from afforestation or reforestation can lower the albedo, which will offset the carbon storage to some extent. Forest trees in general also emit volatile organic compounds such as isoprene, which have a climate warming effect [47]. As discussed earlier, preserving biodiversity in all ecosystems is also vital, and may be negatively affected by AFOLU changes in general. The critics also argue that important factors such as rising demand for food production—the UN expect the global population to continue to rise until the end of the twenty-first century in their medium forecast [60]—will greatly cut biological CDR potential. Smith and Torn [52] examined the ecological consequences of CDR from either afforestation or bioenergy carbon capture and sequestration (BECCS). They concluded that even one Gt C/year would cause 'major perturbation to land, water, nitrogen, and phosphorus stocks and flows'.

A more fundamental problem for afforestation or reforestation is whether it offers more than a temporary solution in a world of rapid ongoing climate change [1, 45]. Pereira and Viola [45] think that forest loss alone could produce catastrophic climate change, while Denning [17] points out that the Eastern Amazon region has ceased to be a carbon sink. Extreme weather is on the rise, and with it extended droughts and heatwaves. These lead to a rise in forest tree mortality through pest damage and wildfires, which have recently caused widespread destruction in Australian, Canadian, Russian, and US forests. Vaughan [61] even entitled his 2019 *New Scientist* article 'Dawn of the Pyrocene'. Walker et al. [64] point out that such fires in boreal regions can lead to the release of stored carbon in forest soils.

2.4.2 Mechanical Sequestration

A variety of mechanical methods for removing CO_2 from the atmosphere have been proposed, and one—carbon capture and storage (CCS)—is already practiced on a small scale. But the quantity of CO_2 so sequestered—around 10 million tonnes yearly—is dwarfed by the tens of billions of tonnes needed. CCS would capture CO_2 from the exhaust stacks of FF power stations in amine solutions, separate, and compress the CO_2, then, after transport to the storage site, bury it deep underground. (Some CO_2 is already captured and sent by pipeline to oilfields, where it is sold for use in enhancing recovery of oil from depleting fields. The CO_2 used is thus sequestered underground. However, using CO_2 to extract even higher amounts of FFs will not help mitigate CC.)

Another approach is Direct Air Capture (DAC). One proposal would employ many thousands of devices to trap CO_2 from ambient air in liquid solvents. A Swiss company, Climeworks, in 2021 opened a 4000 tonne/year plant in Iceland, called Orca, which is powered by geothermal energy [22]. However, the cost (1000 Euro per tonne CO_2) appears prohibitive. Capture from exhaust gases of all large FF plants would only sequester less than half of CO_2 from all FF combustion, since it is not feasible to capture CO_2 from small individual emitters such as transport vehicles. Further, if FF use finally declined for CC mitigation reasons, the annual amount for capture would also decline. DAC faces no such limitation; GHG emissions from all sources and past years could be captured. Countries could contract to remove CO_2 from the global atmospheric commons for other countries for a fee, and countries with historically large cumulative emissions could theoretically use DAC to address the carbon equity issue. Long-distance pipeline transport of CO_2 could be avoided by siting the DAC devices near suitable underground storage sites.

Both methods are energy-intensive, particularly DAC [10]. For DAC, extracting CO_2 from ambient air with around 0.04% is much more energy-intensive than extracting CO_2 from a power station exhaust gas stream, with concentrations of perhaps 10_15%. Energy costs are far lower for CCS, provided power stations are optimised for CO_2 capture, but very few existing FF plants are. While DAC energy intensity estimates vary, capturing current annual CO_2 emissions from the air could require anywhere from 30 to 50% of the current annual global energy supply. But whether CCS, DAC, or BECCS is implemented, they all share the need for deep geological disposal of the captured CO_2. This disposal has not been tried at the many gigatonne levels needed for significant CC mitigation. The technique also faces several drawbacks, including caprock integrity, and groundwater contamination. Deep injection of CO_2 can cause seismic events, and earthquakes can damage caprock integrity [36].

A very different approach for extracting CO_2 from the air is enhanced weathering (EW) of basic rocks, such as dunnite and harzburgite, which both contain large percentages of olivine, and basalt, commonly found on Earth. The rocks would be ground to a fine powder and spread over land to absorb CO_2 [19, 56]. In some situations, spreading the fine powder could neutralise soil acidity, and so enhance agricultural productivity. Beerling et al. [5] have argued that four countries with large surface areas (China, India, the USA, and Brazil) could annually sequester from 0.5 to 2 Gt of CO_2 'with extraction costs of approximately US$80–180 per tonne of CO_2'.

The proposal by Taylor et al. [56] is far more ambitious. They claimed that EW could potentially 'lower atmospheric CO_2 by 30–300 ppm by 2100'. The costs for achieving a 50 ppm reduction were estimated at '\$60–600 trillion for mining, grinding, and transportation, assuming no technological innovation, with similar associated additional costs for distribution'. The cost variation is mainly the result of different annual pulverised silicate rock application rates and the rock composition; they assumed one and five kg/m^2 over 20 million km^2 of the tropics, the region of most rapid weathering. The quantities involved would thus be immense; up to 100 Gt each year. But as the authors themselves discussed, the 'distribution of pulverised rock carries health risks to anyone coming in contact with it because the particle sizes involved are respirable'. Ecosystems would also be at risk [19]. As for CC itself, the health costs of EW would be borne by tropical countries, even if the projects were financed by high-emitting countries.

2.4.3 Discussion

Like Solar Radiation Management (SRM), discussed below, significant deployment of CDR (possibly using a mix of methods) would enable the continued use of FFs. As such, neither method can be considered a permanent solution, because sooner or later, FF energy depletion—particularly for oil—would become an issue. At present, the worry for FF companies and oil-exporting countries is that they will be left with stranded assets [11], as the world cuts back on FF consumption. 'Peak oil' *production* would never happen; instead, we would have peak oil *consumption*. Above, we showed that heavy deployment of CCS and especially DAC would not only allow FF use to continue, but would require even greater energy use than today for a given level of delivered energy. Hence, if either of these approaches were adopted, FF depletion, particularly oil depletion, would once again be of concern.

Bentley et al. [6] have shown that for conventional oil sources, 'the plateau in the global production of this oil since 2005 has been resource-limited, at least for oil prices well in excess of \$100/bbl'. Non-conventional oil—from the deep ocean, from Arctic regions, or from tar sands or heavy oil—all have higher monetary and environmental costs per litre of delivered refined product compared with conventional oil. They concluded that the 'supply of low-cost oil is in decline'. Ugo Bardi [3] has claimed in his article 'Peak oil, 20 years later: Failed prediction or useful insight?' that the peak oil idea failed to gain traction because it was 'considered incompatible with the commonly held views that see economic growth as always necessary and desirable and depletion/pollution as marginal phenomena that can be overcome by means of technological progress'.

The conclusion is that neither biological nor mechanical methods for CO$_2$ removal—in the increasingly unlikely event that they are implemented at large scale—can ever be more than temporary solutions for our climate change-related energy problems. Both Anderson and Peters [2], and Larkin et al. [31] doubt that negative emission technologies will work at the scale required, as they are largely unproved. Smith et al. [54] are somewhat more optimistic about the prospects for

NETs, but conclude that 'there are risks associated with relying heavily on any technology that has adverse impacts on other aspects of regional or planetary sustainability'. A further limitation on CDR reductions comes from a CO_2 feedback effect. Just as present CO_2 emissions to the atmosphere are partly stored in soil and ocean sinks, CO_2 withdrawal from the atmosphere will be counterbalanced by CO_2 additions from these same sinks, partly negating the effectiveness of CDR, but also reducing ocean acidification [24]. CDR methods have been discussed for decades; the lack of implementation does not bode well for their future.

2.5 Solar Radiation Management (SRM)

Another proposed—and controversial—technical fix is geoengineering. The idea was first suggested more than four decades ago, but renewed interest followed the publication of a paper by Nobel Laureate Paul Crutzen on the topic in 2006, prompted by the lack of progress in conventional climate mitigation methods [13]. Its most discussed form is solar radiation management (SRM), in which millions of tonnes of sulphate aerosols would be placed and maintained in the lower stratosphere. The technique would mimic the effect of large volcanoes, such as the Mount Pinatubo eruption in 1991, which sent an estimated 10 Mt of sulphates into the lower tropical stratosphere and led to a global cooling, which at its peak reached around 0.5 °C [58]. The aerosols would block some insolation, thus lowering (daytime) temperatures. But, unlike CDR, with SRM no attempt would be made to lower atmospheric CO_2 concentrations. Progressive acidification of the oceans would therefore continue unchecked—and with it, possible serious ocean ecosystem changes. Two important advantages of SRM are that it could be rapidly implemented—and also stopped rapidly if unforeseen problems arose—and it appears to be far cheaper than alternative approaches to CC mitigation [53].

A variety of other SRM methods have been proposed, and are shown schematically in Fig. 2.4. As can be seen, nearly all involve modifications that would increase Earth's *albedo*—the percentage of incoming solar radiation directly reflected back into space. Most of these involve modification to the atmosphere or stratosphere, but others aim to raise the albedo of Earth's surface, by the use of white surface coatings on roofs and streets, covering vast areas of desert lands with reflective sheets, and even changing vegetation and crops for others with higher albedo. Several methods might be used in concert. An ambitious—and costly—proposal would see the deployment of thousands of mirrors in outer space, which would act as a sunscreen, reducing insolation at the top of the atmosphere. A different approach, cirrus cloud thinning, would enable more of Earth's longwave radiation to escape into space.

Earlier, SRM using sulphate aerosols was envisaged as a means to totally cancel out the radiation forcing from all greenhouse gases (GHGs) including those from both energy use and land use changes such as net deforestation. A constant load of aerosols would be maintained by annual additions, as the aerosols would be removed from the stratosphere by rainout and other processes in a year or so. Schneider and colleagues

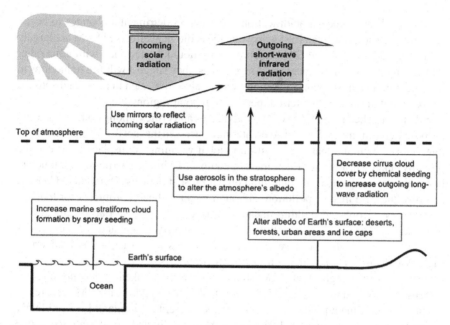

Fig. 2.4 Schematic diagram of geoengineering options for mitigating global warming. Modified from Fig. 9.1 in [36]

[51] have questioned whether sustained global-level geoengineering with sulphate aerosols will work as intended. They argue that rising CO_2 levels would reduce cloud cover by 'modulating the longwave radiative cooling within the atmosphere', paradoxically leading to global warming. Kleinschmitt et al. [28] suggest another possible difficulty: at the higher levels of sulphate aerosols needed for full mitigation, the 'forcing efficiency' of aerosols may be drastically reduced.

For these and other reasons, more recent writings on the topic foresee SRM only partly offsetting anthropogenic climate forcing [27], with other methods being needed as well. The main reason for this change is that SRM would not only lower global temperatures but could also potentially lower precipitation, possibly in key agricultural regions [68]. Methods such as painting urban roofs and roads with reflective paint would have a local cooling effect, but would be insignificant from a global CC mitigation viewpoint. (One key advantage would be that the city or region implementing such a program would reap the benefit; such is not the case for conventional CC mitigation methods, thus avoiding the free-rider effect.) More ambitious—and very costly—proposals would pump vast quantities of seawater onto Antarctic ice to prevent ice mass loss [62]. A similar proposal would use vast numbers of wind turbines to pump water onto winter sea ice in the Arctic in order to increase ice mass and prevent albedo decline [23].

One line of criticism of SRM concerns ecological effects. If for any reason, SRM was abruptly terminated, the rapid rebound from suppressed global warming would have deleterious effects on ecosystems [59]. One known effect of SRM with continued FF use is ocean acidification, which risks detrimental changes to ocean ecosystems. Zarnetske et al. [68] worry that the ecological effects of SRM are largely unknown, and suggest that SRM interventions should be designed with aims other than average temperature decrease. As examples, they suggest wildfire reductions in targeted regions, preserving ecosystems, and as also advocated in [23, 62] discussed above, maintaining cold winter temperatures in polar regions.

SRM as a policy response to CC has advocates and critics among both science and the policy communities. Several advocates have stressed the fact that comparing SRM intervention with an ideal world with no global warming is misplaced. Instead, its possibly unknown, adverse effects must be compared with the also possibly unknown, but serious consequences of continuing on our present path, as discussed in Chap. 1. It is also the case that air pollution, although it causes health problems for humans and damages crops and building facades, also acts to increase Earth's albedo. As discussed in Sect. 1.2, as air pollution is progressively eliminated for the reasons stated—and, hopefully, because of reductions in fossil fuel use—this inadvertent SRM will decline, leading to additional temperature rises.

Even so, some argue that SRM should not be undertaken under any circumstances. Stephens and Surprise [55] claim that SRM—and even SRM research—would deliver climate policy over to a technocratic elite. Biermann [7] has bluntly warned of the risks involved in deploying SRM in our divided world, stating that 'the current world order is unfit to devise and implement such far reaching agreements on planetary management.' Given that climate engineering has already been used for military purposes [36], and that any SRM intervention, even if only implemented on a regional scale, would produce both winners and losers among nations, the political consequences must be taken very seriously.

2.6 Conclusions and Discussion

For various reasons, the solutions discussed in this chapter for avoiding serious climate change either will be of minor benefit, or have significant unwanted side effects. The first two methods discussed, massive reliance on nuclear power, and major improvements to energy efficiency, have the advantage of reducing reliance on FFs. But, as shown in Chap. 1, only a limited time frame is available for solutions to the risk of catastrophic CC, since the adverse consequences of climate change are already apparent. For example, because the global reactor fleet is ageing, and because of long lead times for new plant construction, it would take many decades for nuclear power to achieve a major share even for electricity production, let alone all primary energy.

SRM is the only mitigation method that could be rapidly implemented, and appears to have far lower energy and monetary costs than CDR methods, but, like CDR, would after a few decades lead to depletion of economically viable fossil fuels. Both CDR and SRM approaches would merely delay the inevitable need for alternative fuels. But what if, in a decade or two, we find ourselves with an impending climate catastrophe?

The various CDR methods all need perhaps decades for large-scale implementation, but the same is not true for SRM. Proponents will argue that whatever the risks of SRM, they are less than continuing on as before. So far, the successive IPCC reports have discussed SRM, but have not included it in their scenarios. Although the UK Royal Society had in 2009 given qualified approval for geoengineering, the US National Academies in their 2021 report [43] now urge a coordinated approach to geoengineering research and governance to determine whether it is a feasible solution. Given the time to implement the other tech fix options discussed in this chapter, will this opposition continue?

References

1. Anderegg WRL, Trugman AT, Badgley G et al (2020) Climate-driven risks to the climate mitigation potential of forests. Science 368:1327
2. Anderson K, Peters G (2016) The trouble with negative emissions. Science 354:182–183
3. Bardi U (2019) Peak oil, 20 years later: Failed prediction or useful insight? Energy Res Soc Sci 48:257–261
4. Bastin J-F, Finegold Y, Garcia C et al (2019) The global tree restoration potential. Science 365:76–79
5. Beerling DJ, Kantzas EP, Lomas MR et al (2020) Potential for large-scale CO_2 removal via enhanced rock weathering with croplands. Nature 583:243–248
6. Bentley RW, Mushalik M, Wang J (2020) Biophys Econ Sustain 5: 10 https://doi.org/10.1007/s41247-020-00076-1
7. Biermann F (2021) It is dangerous to normalize solar geoengineering. Nature 595:30
8. BP (2021) BP statistical review of world energy 2021: 70th edn. London, BP
9. Bradshaw AM, Hamacher T, Fischer U (2011) Is nuclear fusion a sustainable energy form? Fusion Engg Design 86:2770–2773
10. Chatterjee S, Huang K-W (2020) Unrealistic energy and materials requirement for direct air capture in deep mitigation pathways. Nature Comm 11: b3287 https://doi.org/10.1038/s41467-020-17203-7
11. Chevallier J, Goutte S, Ji Q, Guesmi K (2021) Green finance and the restructuring of the oil-gas-coal business model under carbon asset stranding constraints. Energy Pol 149: 112055
12. Claessens M (2020) ITER: The Giant Fusion Reactor. Springer-Nature Switzerland AG
13. Crutzen PJ (2006) Albedo enhancement by stratospheric sulfur injections: A contribution to resolve a policy dilemma? Clim Change 77:211–219
14. Daly H (2020) A note in defense of the concept of natural capital. Ecosys Serv 41: 101051
15. Davies S (2004) The Great Horse-Manure Crisis of 1894. Fdn for Econ Education. https://fee.org/articles/the-great-horse-manure-crisis-of-1894/
16. Davis SC, Boundy RG (2020) Transportation energy data book: Edition 38 (cta.ornl.gov/data)
17. Denning S (2021) Southeast Amazonia is no longer a carbon sink. Nature 595:354–355
18. Fouquet R (2014) Long-run demand for energy services: Income and price elasticities over two hundred years. Rev Environ Econ Pol 8:186–207
19. Goll DS, Ciais P, Amann T et al (2021) Potential CO_2 removal from enhanced weathering by ecosystem responses to powdered rock. Nature Geosci 14:545–549
20. Griscom BW, Adams J, Ellis PW et al (2017) Natural climate solutions. PNAS 114(44):11645–11650
21. Hassanein A, Sizyuk V (2021) Potential design problems for ITER fusion device. Sci Repts 11:2069
22. Hook L (2021) World's biggest 'direct air capture' plant starts pulling in CO2. https://www.ft.com/content/8a942e30-0428-4567-8a6c-dc704ba3460a

23. Hooper R (2019) Arctic rescue squad. New Sci 31: 38–41
24. Intergovernmental Panel on Climate Change (IPCC) (2021) Climate change 2021: The physical science basis. Contribution of AR6, WG1. CUP, Cambridge UK
25. International Energy Agency (IEA) (2020) Key world energy statistics 2020. IEA/OECD, Paris
26. International Atomic Energy Agency (IAEA) (2020) Energy, electricity and nuclear power estimates for the period up to 2050. IAEA, Vienna, Austria
27. Irvine P, Emanuel K, He J et al (2019) Halving warming with idealized solar geoengineering moderates key climate hazards. Nature Clim Change 9:295–299
28. Kleinschmitt C, Boucher O, Platt U (2018) Sensitivity of the radiative forcing by stratospheric sulfur geoengineering to the amount and strategy of the SO_2 injection studied with the LMDZ-S3A model. Atmos Chem Phys 18:2769–2786
29. Krymm R, Woite G (1976) Estimates of future demand for uranium and nuclear fuel cycle services. IAEA Bull 18: 5/6
30. Kurzweil R (2005) Human 2.0. New Sci 187:32–37
31. Larkin A, Kuriakose J, Sharmina M, et al (2017) What if negative emission technologies fail at scale? Implications of the Paris Agreement for big emitting nations. Clim Pol. https://doi.org/10.1080/14693062.2017.1346498
32. Lu D (2021) Bitcoin in China has a polluting future. New Sci 17 April: 19
33. Marchetti C (2009) On energy systems historically and in the next centuries. Glob Bioeth 22(1–4):53–65
34. Moriarty P (2021) Global nuclear energy: an uncertain future. AIMS Energy 9(5):1027–1042. https://doi.org/10.3934/energy.2021047
35. Moriarty P (2021) Global transport energy. Encyclopedia 1: 189–197. https://doi.org/10.3390/encyclopedia1010018
36. Moriarty P, Honnery D (2011) Rise and fall of the carbon civilisation. Springer, London
37. Moriarty P, Honnery D (2015) Reliance on technical solutions to environmental problems: Caution is needed. Environ Sci Technol 49:5255–5256
38. Moriarty P, Honnery D (2017) Sustainable energy resources: prospects and policy. Chapter 1 in Rasul MG, et al (eds) Clean energy for sustainable development, Academic Press/Elsevier, London. pp 3–26
39. Moriarty P, Honnery D (2019) Energy efficiency or conservation for mitigating climate change? Energies 12:2019
40. Moriarty P, Honnery D (2020) New approaches for ecological and social sustainability in a post-pandemic world. World 1:191–204. https://doi.org/10.3390/world1030014
41. Moriarty P, Honnery D (2021) Reducing energy in transport, building and agriculture through social efficiency. In: Lackner M, Sajjadi B, Chen, W-Y (eds) Handbook of climate change mitigation and adaptation, 3/e. Springer, NY
42. National Oceanic and Atmospheric Administration (NOAA) (2021) Trends in atmospheric carbon dioxide (https://www.esrl.noaa.gov/gmd/ccgg/trends/)
43. National Academies of Science Engineering Medicine (2021) Reflecting sunlight: recommendations for solar geoengineering research and research governance. Consensus Study Report. https://www.nap.edu/resource/25762/Reflecting%20Sunlight%204-Pager.pdf
44. Organization of the Petroleum Exporting Countries (OPEC) (2020) OPEC World Oil Outlook 2020. http://www.opec.org
45. Pereira JC, Viola E (2018) Catastrophic climate change and forest tipping points: blind spots in international politics and policy. Glob Policy 9:513–524
46. Petsko GA (2007) They fought the law and the law won. Genome Biol 8(10):111. https://doi.org/10.1186/gb-2007-8-10-111
47. Popkin G (2019) The forest question. Nature 565:280–282
48. Post R (1971) Fusion power. PNAS 68(8):1931–1937
49. Ramana MV (2021) Small modular and advanced nuclear reactors: a reality check. IEEE Access 9:42091
50. Ritchie H, Roser M (2018) Plastic Pollution. OurWorldInData.org. https://ourworldindata.org/plastic-pollution

51. Schneider T, Kaul CM, Pressel KG (2020) Solar geoengineering may not prevent strong warming from direct effects of CO_2 on stratocumulus cloud cover. PNAS 117:30179–30185
52. Smith LJ, Torn MS (2013) Ecological limits to terrestrial biological carbon dioxide removal. Clim Change 118:89–103
53. Smith JP, Dykema JA, Keith DW (2018) Production of sulfates onboard an aircraft: implications for the cost and feasibility of stratospheric solar geoengineering. Earth Space Sci 5: 150–162 (https://doi.org/10.1002/2018EA000370)
54. Smith P, Davis SJ, Creutzig F et al (2016) Biophysical and economic limits to negative CO_2 emissions. Nature Clim Change 6:42–50
55. Stephens JC, Surprise K (2020) The hidden injustices of advancing solar geoengineering research. Global Sustain 3(e2):1–6
56. Taylor LL, Quirk J, Thorley RMS et al (2016) Enhanced weathering strategies for stabilizing climate and averting ocean acidification. Nat Clim Change 6:402–406
57. Tenner E (1996) Why things bite back: technology and the revenge effect. Alfred A Knopf, NY
58. Trenberth KE, Dai A (2007) Effects of Mount Pinatubo volcanic eruption on the hydrological cycle as an analog of geoengineering. Geophys Res Lett 34(L15702):1–5. https://doi.org/10.1029/2007GL030524
59. Trisos CH, Amatulli G, Gurevitch J et al (2018) Potentially dangerous consequences for biodiversity of solar geoengineering implementation and termination. Nature Ecol Evol 2:475–482
60. United Nations (UN) (2019) World Population Prospects 2019. https://population.un.org/wpp/
61. Vaughan A (2019) Dawn of the pyrocene. New Sci 27 July: 20–21
62. Vaughan A (2019) Drastic geoengineering could help stem rising seas. New Sci 27 July: 16.
63. Veldman JW, Aleman JC, Alvarado ST, et al (2019) Comment on The global tree restoration potential. Science 366: eaay7976
64. Walker XJ, Baltzer JL, Cumming SG et al (2019) Increasing wildfires threaten historic carbon sink of boreal forest soils. Nature 572:520–523
65. Wikipedia (2021) Thomas Midgley Jr https://en.wikipedia.org/wiki/Thomas_Midgley_Jr
66. Wikipedia (2021) List of nuclear power accidents by country https://en.wikipedia.org/wiki/List_of_nuclear_power_accidents_by_country
67. Wikipedia (2021) Unintended consequences https://en.wikipedia.org/wiki/Unintended_consequences
68. Zarnetske PL, Gurevitch J, Franklin J, et al (2021) Potential ecological impacts of climate intervention by reflecting sunlight to cool Earth. PNAS 118: e1921854118

Chapter 3
The Limits of Renewable Energy

Abstract The ability of RE to fully supply all global energy needs, both now and in the future, is subject to considerable controversy. This chapter first examines the concept of Energy Return on Investment (EROI) as a crucial test for FF project feasibility, and so the technical potential of RE sources. It is found that inclusion of the full input costs of RE, including the energy costs of maintaining ecosystem functions, leads to large reductions in evaluated EROI, even though exact values are not available. In principle, a graph of EROI versus energy production can be drawn for an energy source, enabling technical potential to be determined. The chapter then reviews in turn the prospects for solar, wind, biomass, hydro, geothermal, and ocean energy, pointing out their benefits as well as their environmental disadvantages. Although bioenergy and hydro are the most important sources today, wind and solar energy have by far the greatest potential—and are the most rapidly growing. It is concluded that renewable energy may not be as green or abundant as often portrayed.

Keywords Bioenergy · Energy input costs · Energy return on investment · Geothermal energy · Hydropower · Ocean energy · Solar energy · Wind energy

3.1 Introduction to Renewable Energy

The future of RE is the subject of increasing debate. Some researchers think that RE sources can not only fully substitute for FFs but can also fully avoid CO_2 energy emissions. Others are sceptical of this claim, viewing RE at best having a more modest role to play in CC mitigation, particularly given the short time frame remaining for effective mitigation [42, 47, 60]. RE was a more feasible solution in 1990, the year of the first IPCC report than it is today, over three decades later, given both the rapid shrinkage of the size of the remaining 'carbon pie' over this period, the increasing quantity of annual FF use that RE must replace, and the limited time available. Yet a massive switch to RE sources is the cornerstone of the so-called 'Green Economy', which, it is promised, will enable—or even accelerate—economic growth, in an ecologically sustainable way.

In this chapter, we examine the prospects for RE, starting in Sect. 3.2 with the critically important concept of energy return on investment (EROI). Section 3.3 then

P. Moriarty and D. Honnery, *Switching Off*, Energy Analysis,
https://doi.org/10.1007/978-981-19-0767-8_3

discusses an increasingly major input into RE production: the energy needed to mine and refine the materials needed, and the pollution this mining inevitably causes. In Sect. 3.4, we look at the potential, and advantages and disadvantages of each of the most commonly used—or advocated—RE sources. Section 3.5 offers some tentative conclusions on RE prospects.

3.2   The Importance of Energy Return on Investment (EROI)

How do you decide whether to undertake a project, whether public, commercial, or even personal? The French civil engineer, Jules Dupuit, in 1848 developed the idea of cost–benefit analysis (CBA) to determine 'the social profitability of a project like the construction of a road or bridge' [74]. The CBA ratio compares the benefits of the project over its life, compared with the cost of construction and operation. For example, a new bridge which provides a more direct route than was formerly available will result in savings of both time and money for travellers. Both the benefits and costs need to be expressed in monetary terms for the ratio to be calculated. The net present value of both benefits and costs is determined by the use of a social discount rate. Nevertheless, the idea of qualitatively weighing up the benefits and disadvantages of any action has been around for millennia.

Energy analysis, or energy return on investment (EROI), is based on a similar idea—although one important difference is that future energy costs or benefits are not discounted when expressed in energy terms. Consider an electric power plant. The common-sense idea is that all the energy inputs for building this energy conversion system, such as a wind farm or a coal-fired power station, should be less than the energy delivered by the energy plant over the course of its operating life. The energy inputs should include the energy costs to manufacture and construct the energy conversion device; the operation and maintenance energy costs; the transport energy costs for the components; the energy costs of any additions needed to the transmission system; and any energy costs for decommissioning, removal, and site remediation. Charles Hall, a pioneer in energy analysis, argues that the importance of EROI extends well beyond energy, to biology, economics and sustainability in general [20, 21].

Like CBA analysis, a proper determination of the correct value for EROI is not a simple task. For a start, there are two different approaches to EROI calculation: input–output analysis and process-based. The first approach infers energy costs from the average energy intensity of various economic sectors, while the second sums the energy costs of the various stages in the manufacture and construction of energy systems. A second, and more important, difficulty is that there are various ways of assessing EROI, even for a given basic approach. Although developed a decade ago by Murphy et al [50], we will use the more recent formulation of De Castro and Capellán-Pérez [13]. The latter authors list three methods, which they term *standard*, *point of use*, and *extended energy return on energy invested*. These three methods they define by the following three equations:

$$Method\ 1\ (standard) = \frac{\text{Energy delivered by the plant or energy source}}{\text{Energy used to deliver the energy by the energy system}}$$

$$Method\ 2\ (point\ of\ use) = \frac{\text{Final energy delivered to the final consumer}}{\text{Direct final energy used to deliver energy}}$$

$$Method\ 3\ (extended) = \frac{\text{Final energy delivered to the final consumer}}{\text{Direct + Indirect final enegy used to deliver the energy}}$$

We have earlier called the indirect costs in Method 3, the ecosystem maintenance energy (ESME) costs, and the EROI determined from their inclusion, the green EROI or EROIg [42, 45]. ESME can be interpreted as the energy needed to mitigate the damage done to the Earth's ecosystems by an energy system or technology.

For a given energy, their calculated EROI value progressively falls from the first to the third method. For example, for onshore wind energy, the authors give EROI values for the first, second, and third methods as 13.2, 5.8, and 2.9 respectively. While it is very clear EROI depends on the method used, it is also clear that the nearer energy is to the consumer, and the greater is the account given to the cost of production, the smaller the net energy delivered will be. To be a viable energy source, the EROI must evidently be greater than unity, otherwise no net energy is produced. Some researchers argue that it must be significantly greater than unity—their EROI estimates needed for the functioning of industrial society range from 3 to 11 or even higher [18, 22, 32, 47]. Even if methods for EROI calculation were standardised, there would be no unique value for any one energy source; EROI values will not only vary from country to country but also from project to project within a country, depending, for example, on average wind speeds, average insolation levels, or tidal variation at a given site, and the specifics of the supply chain. Nevertheless, Hall [21] has argued that with proper attention to definitions and methodology, these problems can be largely overcome. Figure 3.1 shows schematically one possible relationship between EROI and the cumulative output energy summed from highest EROI to lowest EROI for a particular RE energy source. Potential, often termed technical potential, is the energy that could be produced excluding constraints on location and energy inputs. As the distribution suggests, the most abundant resources are typically found in the low- to mid-range of the EROI distribution; it is here that the cumulative energy rises the fastest.

For fixed energy input, such as the energy required to construct and commission a wind turbine, EROI will largely depend on the quality of the resource, the wind strength. Given the spatial distribution of the resource and its quality, the RE source's capacity to deliver energy can be assessed as a function of location and time. For each EROI method, the sensitivity to the point of use and inputs can then be assessed at regional, national, or global levels. Once a suitable cut-off value for EROI is chosen, the source's true capacity to deliver net energy is revealed. As an example, for wind, the use of EROI enables exclusion of regions where the wind resource

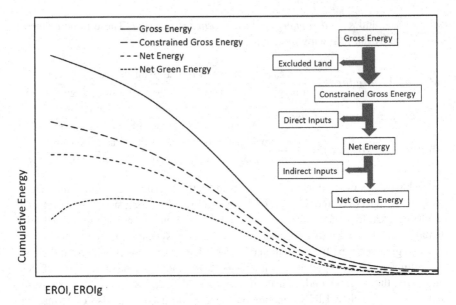

Fig. 3.1 Schematic diagram of cumulative energy against EROI and EROIg for an energy source. Cumulative energy is summed from highest to lowest EROI (from right to left). *Gross Energy* is the cumulative gross energy of the energy source and is related to the source's Earth energy flow; *Constrained gross energy* is cumulative gross energy less resources located on land with physical or social constraints; *Net Energy* is the cumulative gross energy less direct inputs; and *Net Green Energy* is the cumulative gross energy less direct and indirect (ESME) inputs. Further losses will occur as energy is supplied to the consumer

is insufficient to return net energy over the lifecycle of the wind turbine, typically 25 years. Although not directly related to EROI, and independent of resource quality, further exclusions could be made on the basis of accessibility, geography, existing land use, and ecosystem sensitivity. In Fig. 3.1, excluding regions for physical and social reasons lead to the constrained cumulative gross energy, which once energy system inputs are included, falls further to become the cumulative net energy for the resource.

Since it is generally the case that the most accessible resource-intensive sites are usually exploited first, actual average EROI values for a given RE source can be expected to decline over time as regions with lower resource quality are taken up. Indirect energy inputs could be expected to rise through, for example, biodiversity loss as well as other ecological and social damages as land take-up increases. In the early phase of development, technology learning rate can be expected to reduce direct costs, but as the technology matures, gains will become smaller.

There is already indirect evidence that EROI is starting to fall for hydro and geothermal energy developments, at least in some countries [45]. A final problem concerns the exact value of EROI, conventional or otherwise, since we do not have—and may never have—reliable data on ESME costs. For instance, large-scale RE installations can cause a loss of biodiversity [55]. How do we calculate the

ESME costs for biodiversity loss? Such costs will even have an unavoidable ethical component.

For solar and wind RE systems, almost all energy inputs and their impacts occur during the manufacture of the RE system rather than during operation, whereas for FF powered systems the opposite is true; most impacts result from operation [53]. As a consequence, RE system lifetime ESME costs will depend on the mix of energy used in its manufacture, while for fossil fuel systems, ESME largely depends on the fuel used during operation. This adds a further complication for EROI calculations, particularly for RE systems, since our current energy mix is dominated by FFs, but may not be in the future. Although difficult to determine, including ESME costs in the hypothetical example shown in Fig. 3.1 then gives the cumulative net green energy. Cumulative net green energy can be seen to decrease for low values of EROI, as the sum of direct and indirect (ESME) energy inputs exceed output for the individual EROI; net energy additions are then negative.

Because ESME costs such as those for avoiding CO_2 emissions (for example, by using NETs) are presently ignored in most EROI calculations, average conventional (or standard) EROI values for energy systems powered by FFs are much higher than for RE sources. Weißbach et al. [73], using a consistent basis for calculation, found average values for EROI for FF power generation to be an order of magnitude greater than for wind and solar electricity. Hence, an energy subsidy is presently involved. This energy subsidy will gradually decline as RE replaces FF for energy inputs [47]. Because of the unpaid ESME costs of both FF and RE sources, forecasts of RE output growth may be misleading, should these costs be further addressed. A number of analyses on RE future assume steadily declining costs, but do not consider the future need for energy conversion and storage of intermittent RE, nor the economic subsidy from cheap FFs. A price on carbon would favour RE sources, but this advantage will be offset by both the higher costs of FF inputs into RE—and the GHG and other ESME costs of RE sources themselves. The costs of all energy sources would rise.

3.3 Materials for RE: Energy Costs, Pollution, and Scarcity

In this discussion on materials and their ecological costs, we implicitly use the third method of De Castro and Capellán-Pérez, since we believe including indirect energy investments as energy inputs gives the fairest indicator of the real costs of an energy source. As suggested by Moriarty and Honnery [42, 45], including these as ESME costs related to environmental damages requires damages to be converted to an equivalent energy cost. For example, for a mining site that produces the metals necessary for RE device manufacture, ESME can be calculated as the energy cost of restoring the site. That they are often ignored, in full or in part, is evidenced by the numerous tailings dam failures around the world, with their serious ecological damages [54]. These failing dams can result in fast-moving mudflows, each with releases of up to 40 Mt [57]. Cumulatively in 2020, just under 10 billion tonnes of tailings were stored in tailings dams worldwide [3]. An increase in extreme weather conditions, such as flooding, could make tailings dam failure more likely, just as it does with conventional dams. Should a tailings dam fail, given the significantly increased land area needing restoration, the increase in restoration energy costs would see the ESME costs increase for a device that used any minerals extracted from the restored mine.

These material ESME costs are more important for RE sources than for FF or nuclear sources, since the materials input for RE per MW of generating capacity are usually much higher than for other energy sources, as shown in Table 3.1. A single 100 MW wind farm will need around 30,000 tonnes of iron ore, 50,000 tonnes of concrete, and 900 tonnes of plastics for the turbine blades [37]. Hertwich et al. [23] have also shown the high demand for specific materials for some RE sources, particularly copper and aluminium for PV, and steel for wind and STEC. Not only is material usage per MW greater but also, as noted above, for most RE sources, most input energy must be expended before a single kWh can be generated. So, for a wind turbine farm, the turbine must be manufactured and erected on site, access roads built, and connections to the existing grid made before any power can be generated. Only final decommissioning must be deferred to the end of service life. The exception is bioenergy, where like fossil fuel energy, much of the input energy costs consist of that needed for producing and transporting the biomass fuel for combustion as it is needed.

Although a number of rare earths and other elements are important for the production of wind turbines and PV cells, this does not necessarily imply imminent depletion. Jowlett et al. [29] summarise their findings as follows: 'We suggest that environmental, social, and governance factors are likely to be the main source of risk in metal and mineral supply over the coming decades, more so than direct reserve depletion. This could potentially lead to increases in resource conflict and decreases in the conversion of resources to reserves and production.' To avoid dependence on other countries for critical elements, importing countries may decide to open up local deposits, which may have higher direct energy and ecological costs than imported material. And, as an example of environmental and social factors, Nkulu et al. [51]

Table 3.1 Specific materials consumption for various electrical sources, circa 2015, with generation and materials requirements for 2020

Energy source	Materials consumption (kg per MWh)	Current generation (TWh)	Material for current generation (Mt)
Natural gas	1.0		
Hydro	14.0	4297	60.2
Wind	10.2	1591	16.2
Solar PV	16.3	856	14.1
Geothermal	5.2	95	0.5

Source [37]

reported that in DR Congo, people living near the artisanal cobalt mines had high levels of cobalt in their blood and urine, and this cobalt pollution thus represents a significant health hazard. Further, child labour is used for mining. Lower prices come at a cost, but, unlike cleaning up the mining wastes produced, it is not possible even in principle to assign an energy cost to child labour.

Other researchers argue that even ignoring such political considerations, mineral availability could be a constraint on RE production in the coming decades. Moreau et al. [38] have studied 'the availability of metal reserves and resources to build an energy system based exclusively on renewable energy technologies for the year 2050. They concluded: 'The results show that proven reserves and, in specific cases, resources of several metals are insufficient to build a renewable energy system at the predicted level of global energy demand by 2050.' For at least copper, a vital element for all energy production, peak production may even be reached before 2040; globally, copper grades are slowly falling, and are now below 1%. As ore grades fall with progressive extraction, not only does the direct energy to mine and refine each tonne of metal rise—as is already happening with copper—but the tailings volume also rises. For Chile, Rodríguez et al. [56] report that for each tonne of copper concentrate, 151 tonnes of mining wastes are generated, which are often merely deposited directly on the seabed, causing coastal pollution.

Minerals are not the only material inputs needed: another important one is water, which may be needed for cooling in thermal power plants (including those co-fired with biomass such as wood pellets, and solar thermal plants), and for cleaning PV cell arrays and suppressing dust. Dust on PV array or solar thermal energy conversion (STEC) receiver surfaces can cause large declines in electricity output [5]. In future, solar electricity farms may increasingly be located in the high insolation desert regions of the world; if the electricity produced in these regions is converted to hydrogen, freshwater will also be needed for hydrolysis [44], unless linked to large-scale desalination with the added energy costs and environmental impacts [53].

3.4 Renewable Energy (RE): General Considerations

The Earth is a rocky planet orbiting at an average of almost 150 million km from a minor star lying on the main sequence of stars, situated on the outer edge of the disc of the Milky Way galaxy. Despite searching, we are yet to find any similar planet, and should we do so, it is unlikely to be easily reached. In the distant past, during Earth's violent formation around 4.5 billion years ago, the planet was much hotter, because of collisions with planetoids and the presence of radioactive isotopes in the core and crust. Today, the Earth is still slowly cooling, but still emits an average of 0.08 W/m^2 from Earth's surface overall, but only an average of 0.06 W/m^2 from the continental crust. Such heat is the source of geothermal energy, which is estimated to be around 1.3 ZJ per year—one ZJ is equal to 1.0×10^{21} J, or 1000 EJ. This terrestrial average conceals a wide range in output, and the heat rate leakage is far greater in regions of tectonic activity.

The other two sources of energy to Earth are external. The sun's insolation provides
an average of 1366 W at the top of the atmosphere to every square metre of Earth's
surface, which is the ultimate source of wind energy, hydropower, bioenergy, and
wave energy, as well as direct solar energy. It is also the ultimate source of fossil
fuel resources. With an annual value of 5500 ZJ at the top of the atmosphere, this
insolation is orders of magnitude greater than the other two energy sources. The
other external source of energy to the planet is tidal energy, which is tiny compared
with insolation. Its cause is the mutual gravitational attraction between Earth and
its much smaller but close-by moon, and to a lesser extent, between the sun and
the Earth (and the other planets). Tidal energy amounts globally to only about 80
EJ/year. Nearly all of it is dissipated in the oceans, mainly along coastlines, where it
can potentially provide energy. Figure 3.2 schematically represents the Earth energy
flows, while Fig. 3.3 shows how the global output from the five main RE sources
presently used have fared since 1970. To put this into the context of a future energy
system dominated by electricity, we are likely to see RE needing to supply around
100,000–150,000 TWh by 2050, if we are to avoid a significant rise in global surface
temperatures, or up to a 100-fold that is currently provided by wind and solar energy.

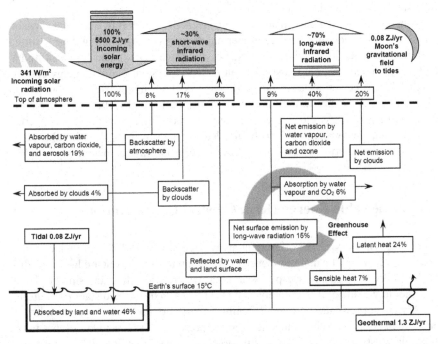

Fig. 3.2 Schematic diagram of annual Earth energy flows. Updated from Fig. 2.1 in [39]

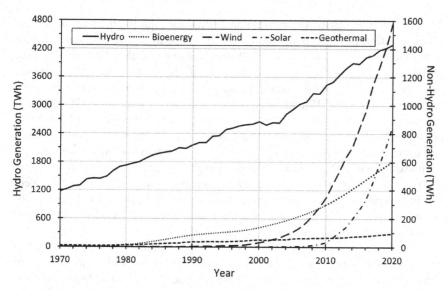

Fig. 3.3 Global hydro, wind, biomass, geothermal, and solar electricity output (TWh) 1970–2020. *Sources* [8, 24, 40, 41]

3.4.1 Solar Energy

Solar energy is vital to all life on Earth. If not for the sun, the average Earth surface temperature would be close to absolute zero, or −273 °C. The sun—and natural global warming from pre-industrial levels of GHG gases—brings the temperature up to about 15 °C. As Fig. 3.3 shows, solar electricity generation was still negligible as late as 2010, and is now the fastest growing energy source. Given that surface of the Earth receives an annual input of around 3,900 ZJ from insolation, resource limitations are not a problem. Solar electricity is produced to some extent in nearly all countries. Most solar electricity is still produced in the OECD, although its share has fallen from over 90% in 2012 to 53% in 2020. The leading producers in order in 2020 were China, the US, Japan, India, and Germany [8].

The main areas of concern centre on two related questions: first, what effect will the need for energy conversion and storage have on solar energy prospects, and second, what is the EROIg for PV cells and STEC plants? The first question arises because solar energy, like wind energy, provides output in the form of electricity only intermittently. Even if direct electrical energy share rises in future, because of, for example, conversion of surface vehicles to electric drive, non-electrical energy will still be needed. The best approach may be to convert electricity in excess of grid needs into hydrogen, and store it as a compressed gas. It can then later be re-converted into electricity using fuel cells, or used directly as a heat source [44]. Each of the steps—conversion, hydrogen storage, and possibly re-conversion to electricity in a hydrogen fuel cell—incurs energy losses.

Even for the calculation of conventional EROI for solar PV, there are huge variations in estimates in the published literature. At the upper extreme, one estimate is that the energy cost of the PV cells themselves (not including balance of system energy costs) will in future be paid back in one day [15]. At the other extreme, there are claims that for the levels of insolation found in Germany and Switzerland, there is no net energy [17], although other researchers dispute this [47]. Although impressive technology advances in PV cell efficiency are occurring, Blanco et al. [7] have asked whether such progress has been at the expense of environmental sustainability, as the production of PV cells involves significant environmental costs from toxic wastes [60]. Similarly, Van den Bergh et al. [69] have wondered whether our pursuit of solar energy will merely be an exercise in 'environmental problem shifting'. Again, full energy input costs need to be considered for all candidates for green energy. For STEC, the study of actual plants around the world—all in areas of good insolation—showed very low values of EROI [12].

One form of solar energy which is under-appreciated—and under-counted in energy statistics—is passive solar energy. This energy use, which can be hard to separate from energy conservation, has been practiced for millennia. In hot climates, it may simply involve the use of thick walls, which can store heat during the day and release it to increase night-time indoor temperatures. Today, special materials are being developed, such as phase-change materials and special surface coatings [46]. However, the efficacy of both active and even some forms of passive solar energy can be reduced if SRM is instituted, since it will cause light scattering. This will lower output for STEC and focusing PV arrays; for the STEC plants in Barstow, California, output declines were actually recorded after the Mount Pinatubo eruption [49].

3.4.2 Wind Energy

Wind energy is second only to hydro for global electricity output, but like solar electricity is an intermittent energy source. Nevertheless, reliability can be somewhat improved by a combination of wind and solar. Like solar energy, wind energy is now used in many countries, with the five leading nations for wind use in 2020 being China, the US, Germany, India, and the UK [8]. An advantage of onshore wind energy is that actual land use is small per MW installed, since most of the area of a wind farm can still be used for grazing and agriculture. For offshore turbines, there is no land loss at all.

Nevertheless, wind farm developments have been subject to citizen opposition, not only in Europe and the US but also in lower income countries. Such developments often involve a clash between local environmental values and global ones, such as CC mitigation. Voigt et al. [72] have also claimed that wind power development is at the expense of biodiversity—another global problem. Operating turbines do kill millions of birds and bats each year, although bird deaths are much less than those from vehicle and building collisions, and from predation by domestic pets. For the US alone for the year 2012, one estimate is that turbine deaths for birds were 234–573 thousand, and for bats 600–888 thousand [62]. Other local opposition concerned visual pollution, reported, but much disputed, health problems for nearby residents, and even house valuation reductions [48]. Although gross global wind power potential is second only to solar, a number of restrictions act to greatly reduce this gross potential. These include no placement of wind turbines in nature conservation areas, mountainous regions, or near settlements, but the most important restriction on potential is access to an existing grid [30].

Large-scale wind farms, such as will be needed if RE is to replace FF, can even directly affect climate. Tall turbines can change patterns of atmospheric turbulence and surface roughness, as well as slow wind speeds [1]. For the US, Miller and Keith [36] argue that producing all electricity by wind turbines would warm the land surface of the continental US by 0.24 °C. In Germany, slower wind speeds and increased atmospheric turbulence have been measured 'extending 50–75 km downwind of Germany's offshore wind plants' [35].

Annual wind turbine output deteriorates with age. Using the existing 282 wind farms in the UK in 2012 as a case study, Staffell and Green [64] found that 'load factors do decline with age, at a similar rate to other rotating machinery.' Wind turbine output declined at about 1.6% per year, 'the average load factor declining from 28.5% when new to 21% at age 19'. Another factor lowering net output in cold climates is the need to divert some output energy for de-icing in cold spells.

3.4.3 Biomass Energy

Biomass energy has been the mainstay of the human energy system for many millennia, and some think it can regain this position, after our all too brief flirtation—compared with the long reign of bioenergy—with stored fossil fuels. But in 1800, when bioenergy accounted for 95% of all energy, the world population was under one billion. In mid-2021, it approached 7.9 billion, and is projected to reach 9.8 billion in 2050 and 10.9 billion in 2100, according to UN medium variant projections [68]. In 2018, bioenergy accounted for only an estimated 9.3% of global energy, or roughly 55 EJ, but little of this was modern bioenergy as used in power stations or for liquid fuels [8, 26]. What are the prospects for a revival?

For a start, the 'maximum global limit for all human biomass use, whether for food, forage, energy, or materials, is ultimately fixed by the net primary production (NPP) of Earth's terrestrial ecosystems, defined as the gross annual fixation of living plant matter, minus respiration' [43]. For Earth overall, NPP is about 3000 EJ; for the terrestrial area, around 1900 EJ. Far from increasing over time, as might be expected from agriculture yield improvements, global NPP has shrunk an estimated 45% over the past two millennia [59]. Unlike other RE sources, there is significant literature on the human appropriation of net primary production (HANPP), but unfortunately the estimates vary from 10 to 55% of NPP, depending on what items are included. Nevertheless, using a consistent metric, HANPP appears to have doubled over the twentieth century [31].

Several points are important to keep in mind when thinking about bioenergy potential. First, there are limits on HANPP, since NPP must ultimately feed all heterotrophic species, not just humans. For this reason, HANPP is probably limited to no more than 45–47% of NPP (or approximately 900–950 EJ), before ecological deterioration starts to reduce the *absolute* NPP value. Second, HANPP must satisfy *all* human uses of biomass, not just energy. Two other human uses have always been important: biomass for food, and for materials, including construction. The most direct conflict with food production is the diversion of foodstuffs—cereals, sugar, soybeans, and vegetable oils—for transport biofuels, mainly ethanol. Production reached 3.75 EJ in 2019 (about 0.7% of all commercial energy), although production declined in 2020. Production is dominated by the US and Brazil; together they produced 58% of the world total in 2020 [8].

For ethical reasons, it is vital to adequately meet the food needs of all Earth's people, although, as discussed further in Chap. 5, diets can be altered. (However, a complication is that significant amounts of energy are needed for food and biomaterials production.) This competition from different uses also applies to biomass inputs such as water. Stenzel et al. [65] argued in their eponymous paper that 'Irrigation of biomass plantations may globally increase water stress more than climate change.'

The third use for biomass, materials, is also important, because replacing energy-intensive structural materials—such as steel or reinforced concrete—with timber will reduce the energy and GHG costs for a given structure. Nevertheless, construction timber can be combusted for power generation at the end of its service life to provide bioenergy [43].

The combination of these factors means that the biomass available for energy use will progressively decline as the food and material needs of humankind increase. Both food and timber production produce wastes—harvest residues, such as stover from cereals, and branches, bark, and sawdust from forest felling and sawing. Only some of these wastes can be sustainably used for energy; much needs to be left to both maintain soil fertility and prevent wind and water erosion.

Bioenergy is only carbon neutral in the long run, when new biomass has fixed the CO_2 emitted during bioenergy combustion. In Chap. 1, we stressed the limited time for action if the world is to avoid catastrophic CC—and crossing Earth's other sustainability limits as well. Sterman et al. [66] have compared electricity generation from biomass and coal. They point out that using bioenergy initially produces more CO_2 for a given electric power output, because of its lower combustion efficiency. Until that amount of CO_2 released is fixed, the atmosphere will have a higher CO_2 content than would be the case for coal. Regrowth—except for short rotation energy crops—may take decades, and as discussed in Chap. 2, may be at risk from ongoing climate change.

Bioenergy is presently the world's leading RE source, and is produced in nearly all countries. It is unique among RE sources in that it is a chemical energy source, and can be converted into gaseous and liquid forms such as biogas and bioethanol. As dried wood pellets, it is even exported around the world. Estimates in the published literature for global bioenergy potential are plentiful, but—as for other RE sources—show a very large range, from a few tens to many hundreds of EJ. Some estimates even exceed the upper limit for HANPP (900–950 EJ) given earlier. Given that HANPP must be partitioned among the three basic uses, it is not even theoretically possible to estimate its potential—in the end, it depends on the ethics of food distribution.

3.4.4 Hydropower

As Fig. 3.2 shows, hydropower has always been the leading RE electricity source, and in 2020 accounted for 4297 TWh (16%) of world electricity, although wind and solar show higher growth rates. Provided there is sufficient storage, hydro is a fully dispatchable energy source, and is widely available, with a number of countries producing most of their power by hydro [8]. Most studies also show a far higher EROI value for hydro than for other RE sources [e.g. 73], mainly because hydro installations can last for a century, compared with 20–30 years for most other RE systems. Nevertheless, there is some evidence that their global value of EROI has declined in recent decades, since the growth rate in output (TWh) lagged behind the capacity (GW) growth [27, 48]. In 2020, the five leading hydropower producers were China, Brazil, Canada, the US, and Russia [8].

In contrast to other RE sources, published estimates for hydro technical potential do not vary much. Recent estimates put its potential as some 3–4 times its present level [41, 47]. But whether global hydropower can be expanded to this extent is increasingly doubtful because of the serious ecological—and even climate change—problems facing future hydro projects. Zarf et al. [77] have shown that global fresh-water megafauna in regions such as South America and SE Asia is at risk from large hydropower projects. Using the Mekong river as a case study, Williams [75] has discussed the many-faceted damages—both ecological and socio-economic—caused by hydro dam development. Jaramillo and Destouni [28] have pointed to a very different effect of large dams: water evaporation from the large surface area of hydro reservoirs, a problem in an increasingly water-stressed world.

Most remaining hydro potential is in the tropical regions of Africa, South America, and Asia, but hydro development in the tropics can lead to large direct GHG emissions. It happens this way: if vegetation, especially forest areas, is flooded by the reservoir, it will undergo a mixture of aerobic and anaerobic decay, producing CO_2 and CH_4. Also, vegetation on the edges of the reservoir will be subject to a growth/decay cycle as the water level rises and falls. For tropical dams, the GHG emissions can exceed those from an equivalent gas-fired plant over the first 20 years of its life [16]. By comparing emissions before and after dam development, Bertassoli et al. [6] showed that these emissions can still be significant in run-of-river plants. For some hydro reservoirs, a further direct climate forcing effect is caused by the decrease in albedo following the replacement of vegetational land cover by water [76]. Hydro is undoubtedly a renewable energy source, but, as the above discussion shows, whether it should be considered an ecologically (or socially) sustainable source is questionable.

Given that climate is already changing, and that further changes can be expected even if we do implement meaningful policies soon, global and regional precipitation patterns will change further. In general, average precipitation levels will rise in a warming world, with some regions such as northern Europe experiencing increases in hydro potential, but others will experience declines [25]. These declines will occur because higher temperatures also mean more evaporation, and in mountain regions, a change in seasonal streamflow patterns as spring snowmelt of the stored snowpack is advanced. The expected rise in intensive rainfall events in catchment areas will also lead to more reservoir sedimentation and loss of capacity. Land use changes can also affect catchment stream runoff and thus hydro potential. In the Amazon basin, further forest clearance will initially increase runoff because of reduced evapotranspiration from trees. But because this evapotranspiration provides rainfall elsewhere in the basin, if forest cover is cut by 40%, overall hydro potential in the Amazon could be reduced to a mere 25% of its original potential [67].

On 7 February 2021, two hydro dams were severely damaged in the Indian Himalaya, with 200 people killed or missing [61]. The cause was a 'massive rock and ice avalanche'. This disaster illustrates only one of the uncertainties facing hydro development in the Himalayas, an important 'mountain water tower'. For some decades to come, melt from glaciers will increase streamflow and hydro capacity. But eventually, streamflow contribution from glacial melt will dry up as glaciers shrink, and streamflow will become reliant on precipitation, and thus more temporally skewed [52]. Hydro output could fall. In any case, planning of new hydro projects will become more uncertain, given the difficulty of forecasting output a century in advance.

Finally, Briones-Hidrovo and colleagues [9] have attempted to assess what they call 'the net environmental performance of hydropower', considering the adverse ecological and socio-economic effects mentioned earlier. They examined two hydro schemes in Ecuador, a run-of-river hydro plant and a conventional hydro plant. They found that the net environmental performance was negative in both cases, at $0.98/kWh for conventional hydropower and $0.08/kWh for the run-of-river plant. Even without the uncertainty caused by future precipitation and streamflow patterns, it is 'possible that the high conventional EROI for hydro will not be reflected in hydro EROIg values' [48].

3.4.5 Geothermal Energy

Geothermal energy can be used in two ways: for electricity production and for direct heat. Geothermal electricity production is now over a century old, but humans have used geothermal heat for millennia. Since the Earth's core is slowly cooling and the radioactive materials (chiefly various isotopes of thorium, uranium, and potassium) in the crust decaying, the total heat content of Earth, and annual surface heat release, are falling. As is common for other RE sources, estimates for the technical potential of both annual electricity production and direct heat use show huge variations [41].

Total output in 2020 for geothermal energy was 95.1 TWh, or about 0.35% of global energy [24]. For electricity production, although it is generally agreed that production from conventional geothermal fields is limited, some researchers [e.g. 2, 70] see great opportunity for enhanced geothermal systems (EGS) production, because most of the heat content of the terrestrial crust is at depths of a few km. Aghahosseini and Breyer [2] gave a value for the sustainable potential power output of 256 GW (electric) in 2050, which was 16 times the 2020 installed geothermal capacity of about 16 GW [48]. Even this 256 GW value would be <25% of installed global wind/solar electric capacity in 2019. Van der Zwaan and Dalla Longa [70] were even more optimistic, with geothermal electricity output projected to reach from 800 to 1300 TWh annually by 2050, given their assumptions. However, what is lacking in these analyses is any indication of the energy input costs for EGS systems, given the need to sink boreholes to depths of up to 5–10 km. An additional consideration is that much of the geothermal energy is available as low-temperature heat (less than 150 °C), compared to that used in traditional fossil fuel power stations (around 1500 °C), meaning that more power generating equipment is needed per unit of power delivered. Although boilers and related equipment are not needed, as is the case for traditional power stations, the lower overall thermal efficiency resulting from the lower temperatures will typically rule out power generation for much of the available geothermal energy.

A significant amount of geothermal energy that is available is better used for direct heat, for which estimates vary from 14 EJ up to 5000 EJ, and even very much higher [41]. In 2019, total heat energy used was about one EJ, compared with about 280 EJ of terrestrial heat release annually. Hence, very high rates of heat use risk depleting geothermal fields over time—ultimately, they are not sustainable in the long run. Unlike geothermal electricity, which can be transported for long distances, direct heat (in the form of hot water in insulated pipes) cannot be feasibly transported more than a few km. This limits the use of such energy to localities near tectonic activity, which may be thinly populated [48].

A relatively recent innovation is *ground source heat pumps* (called geothermal heat pumps in North America). The latter is a misnomer, since the devices will work in the absence of geothermal heat flows—and they can also be used for *cooling* buildings. They work because of the thermal inertia of the ground (or large bodies of water) over the seasonal cycle, giving an energy sink from which energy can be extracted in winter, or into which heat can be dumped in summer. They are, nevertheless, a useful means for reducing the energy in fossil-fuelled space heating.

At present, the leading five countries for geothermal electricity are the US, Indonesia, Kenya, the Philippines, and Turkey [24]. Geothermal energy in some form is now produced in 88 countries around the world [24], and the electricity produced is dispatchable. Nevertheless, geothermal energy can cause a variety of environmental damages (see, for example, [4, 10, 63]). These 'include various harmful emissions which can impact groundwater quality and human health, seismic tremors, and land subsidence'. An Italian study found that the overall environmental impact of geothermal plants was appreciably higher than for wind and solar energy plants in Italy [4]. Fridriksson et al. [19] found that full fuel cycle CO_2 emissions per kWh produced overlap with those from FF plants, partly because of the CO_2 naturally present in geothermal fluids released during energy production. Of course, as we saw above for hydropower, it is important to allow for natural emissions of all types from geothermal regions in evaluating the *additional* releases attributed to field development.

3.4.6 Ocean Energy

The oceans can potentially produce energy in three ways: wave energy; ocean thermal energy conversion (OTEC); and tidal energy. Tidal and wave converters rely on gravitational or kinetic energy, while OTEC devices exploit the low-temperature differential (up to around 20 °C) between warm tropical surface waters and the cold flows at depth. At this temperature difference, OTEC devices will have even lower thermal efficiency than geothermal systems. Although numerous attempts have been made over the past century or so to exploit both OTEC and wave energy, neither technology at present produces commercial power. In 2020, the only commercially available ocean energy was tidal; it provided only about one TWh of electricity, from 0.53 GW generating capacity.

Most energy researchers do not anticipate ocean energy in any form to be more than a marginal future energy source, even by 2050, although Melikoglu [34] argues that it has the potential to supply all the world's current electricity output. The IEA have projected global annual output rising to around 15 TWh by 2030 [11]. Given that in 2020, the total global electricity output was 26,823 TWh, its share would still be negligible. For tidal energy alone, Van Haren [71] has estimated the maximum theoretical potential at 100 GW, and the practical exploitable potential as much lower. Today, tidal electricity is only produced in a handful of countries, with France and South Korea the leaders. Like other RE sources, new projects for tidal energy can face serious environmental concerns, particularly for estuary barrage-type designs. So, although ocean energy can provide non-intermittent output, it is unlikely to compete with the more established RE sources.

3.5 Conclusions on Renewable Energy

Despite the fact that it is difficult to measure, rough determination of EROI is essential if a RE project is to claim green credentials. For ecological sustainability, and to cover uncertainties in its measurement, EROIg should be significantly greater than unity. In fact, RE projects today are not decided on the basis of EROI calculations. Instead, the comparison of economic costs (with any subsidies factored in) with expected revenues over the project's life is the basis of the decision.

Any energy production method can produce largely hidden or ignored costs for any or all three groups: the non-human natural world; people in countries with low incomes and/or low levels of effective pollution controls; future generations of humans. This chapter has shown the negative impacts of the various RE sources along with their benefits. Nor is this the only challenge facing the evaluation of future RE potential. Others are declining resource quality, the need for energy storage if intermittent RE sources become dominant, and the removal of the energy subsidy from higher (as conventionally assessed) EROI fossil fuels as they are replaced by RE.

References

1. Abbasi SA, Tabassum-Abbasi AT (2016) Impact of wind-energy generation on climate: a rising spectre. Renew Sustain Energy Rev 59:1591–1598
2. Aghahosseini A, Breyer C (2020) From hot rock to useful energy: a global estimate of enhanced geothermal systems potential. Appl Energy 279: 115769.
3. Anon (2020) State of world mine tailings 2020. https://worldminetailingsfailures.org/
4. Basosi R, Bonciani R, Frosali D et al (2020) Life Cycle Analysis of a geothermal power plant: comparison of the environmental performance with other renewable energy systems. Sustain 12:2786. https://doi.org/10.3390/su12072786
5. Bergin MH, Ghoroi C, Dixit D, JJ, et al (2017) Large reductions in solar energy production due to dust and particulate air pollution. Environ Sci Technol Lett 4:339–344
6. Bertassoli DJ Jr, Sawakuchi HO, de Araújo KR, et al (2021) How green can Amazon hydropower be? Net carbon emission from the largest hydropower plant in Amazonia. Sci Adv 7: eabe1470
7. Blanco CF, Cucurachi S, Peijnenburg WJGM et al (2020) Are technological developments improving the environmental sustainability of photovoltaic electricity? Energy Technol 8:1901064
8. BP, (2020) BP statistical review of world energy 2020. BP, London
9. Briones-Hidrovo A, Uche J, Martínez-Gracia A (2020) Determining the net environmental performance of hydropower: a new methodological approach by combining life cycle and ecosystem services assessment. Sci Total Environ 712: 136369
10. Chen S, Zhang Q, Andrews-Speed P, et al. (2020) Quantitative assessment of the environmental risks of geothermal energy: A review. J Environ Mgt 276: 111287
11. Chowdhury MS, Rahman KS, Selvanathan V, et al. (2020) Current trends and prospects of tidal energy technology. Environ, Dev Sustain 23: 8179–8194
12. De Castro C, Capellán-Pérez I (2018) Concentrated solar power: actual performance and foreseeable future in high penetration scenarios of renewable energies. Biophys Econ Resour Qual 3:14
13. De Castro C, Capellán-Pérez I (2020) Standard, point of use, and extended energy return on energy invested (EROI) from comprehensive material requirements of present global wind, solar, and hydro power technologies. Energies 13:3036. https://doi.org/10.3390/en13123036
14. Dorrell J, Lee K (2020) The cost of wind: negative economic effects of global wind energy development. Energies 13:3667. https://doi.org/10.3390/en13143667
15. Espinosa N, Hosel M, Angmo D et al (2012) Solar cells with one-day energy payback for the factories of the future. Energy Environ Sci 5:5117–5132
16. Fearnside PM (2015) Emissions from tropical hydropower and the IPCC. Environ Sci Pol 50:225–239
17. Ferroni F, Hopkirk RJ (2016) Energy Return on Energy Invested (ERoEI) for photovoltaic solar systems in regions of moderate insolation. Energy Pol 94:336–344
18. Fizaine F, Court V (2016) Energy expenditure, economic growth, and the minimum EROI of society. Energy Pol 95:172–186

19. Fridriksson T, Merino AM, Orucu AY, et al (2017) Greenhouse gas emissions from geothermal power production. In: Proceedings 42nd Workshop on Geothermal Reservoir Engineering Stanford University February 13–15, SGP-TR-212

20. Hall CAS (2017) Will EROI be the primary determinant of our economic future? The view of the natural scientist versus the economist. Joule 1: 635–638

21. Hall CAS (2017) Energy return on investment: A unifying principle for biology, economics, and sustainability. Lecture Notes in Energy Volume 36; Springer: Cham, Switzerland

22. Hall CAS, Lambert JG, Balogh SB (2014) EROI of different fuels and the implications for society. Energy Pol 64: (141–152)

23. Hertwich EG, Gibon T, Bouman EA et al (2015) Integrated life-cycle assessment of electricity-supply scenarios confirms global environmental benefit of low-carbon technologies. PNAS 112(20):6277–6282

24. Huttrer GW (2020) Geothermal power generation in the world 2015–2020 update report. Proceeding World Geothermals Congress, Reykjavik, Iceland, April 26–May 2

25. Intergovernmental Panel on Climate Change (IPCC) (2021) Climate change 2021: the physical science basis. Contribution of AR6, WG1. CUP, Cambridge UK (Also earlier reports)

26. International Energy Agency (IEA) (2020) Key world energy statistics 2020. IEA/OECD, Paris

27. International Hydro Association (IHA) (2019) Hydro status report 2019. Available from: https://www.hydropower.org/publications/2019-hydropower-status-report-powerpoint.

28. Jaramillo F, Destouni G (2015) Comment on : Planetary boundaries: Guiding human development on a changing planet.: Science 348:1217

29. Jowitt SM, Mudd GM, Thompson JFH (2020) Future availability of non-renewable metal resources and the influence of environmental, social, and governance conflicts on metal production. Comm Earth Environ 1:13. https://doi.org/10.1038/s43247-020-0011-0

30. Jung C, Schindler D (2021) Distance to power grids and consideration criteria reduce global wind energy potential the most. J Cleaner Prod 317: 128472

31. Krausmann F, Erb K-H, Gingrich S et al (2013) Global human appropriation of net primary production doubled in the 20th century. PNAS 110:10324–10329

32. Lambert JG, Hall CAS, Balogh S, Gupta A, Arnold M (2014) Energy, EROI and quality of life. Energy Policy 64: 153–167

33. Lund JW, Toth AN (2021) Direct utilization of geothermal energy 2020 worldwide review. Geotherm: 101915

34. Melikoglu M (2018) Current status and future of ocean energy sources: a global review. Ocean Eng 148:563–573

35. Miller L (2020) The warmth of wind power. Phys Today 73(8):58–59. https://doi.org/10.1063/PT.3.4553

36. Miller LM, Keith DW (2018) Climatic impacts of wind power. Joule 2:1–15

37. Mills MP (2020) Mines, minerals, and : green energy: a reality check. Manhattan Institute Report, July. http://www.goinggreencanada.ca/green_energy_reality_check.pdf

38. Moreau V, Dos Reis PC, Vuille F (2019) Enough metals? Resource constraints to supply a fully renewable energy system. Resources 8:29

39. Moriarty P, Honnery D (2011) Rise and fall of the carbon civilisation. Springer, London

40. Moriarty P, Honnery D (2012) Preparing for a low-energy future. Futures 44:883–892

41. Moriarty P, Honnery D (2012) What is the global potential for renewable energy? Renew Sustain Energy Rev 16:244–252

42. Moriarty P, Honnery D (2016) Can renewable energy power the future? Energy Policy 93:3–7

43. Moriarty P, Honnery D (2017) Assessing the climate mitigation potential of biomass. AIMS Energy 5(1):20–38

44. Moriarty P, Honnery D (2019) Prospects for hydrogen as a transport fuel. Int J Hydrogen Energy 44:16029–16037

45. Moriarty P, Honnery D (2019) Ecosystem maintenance energy and the need for a green EROI. Energy Pol 131:229–234

46. Moriarty P, Honnery D (2019) Energy accounting for a renewable energy future. Energies 12:4280

47. Moriarty P, Honnery D (2020) Feasibility of a 100% global renewable energy system. Energies 13:5543. https://doi.org/10.3390/en13215543
48. Moriarty P, Honnery D (2021) The limits of renewable energy. AIMS Energy 9(4):812–829
49. Murphy DM (2009) Effect of stratospheric aerosols on direct sunlight and implications for concentrating solar power. Environ Sci Technol 43(8):2784–2786
50. Murphy D, Hall CAS, Dale M, Cleveland C (2011) Order from chaos: A preliminary protocol for determining EROI for fuels. Sustain 3: 1888–1907. https://doi.org/10.3390/su3101888
51. Nkulu CBL, Casas L, Haufroid V et al (2018) Sustainability of artisanal mining of cobalt in DR Congo. Nature Sustain 1:495–504
52. Nie Y, Pritchard HD, Liu Q et al (2021) Glacial change and hydrological implications in the Himalaya and Karakoram. Nature Rev: Earth Environ 2:91–106
53. Palmer G, Roberts A, Hoadley A, et al (2021) Life-cycle greenhouse gas emissions and net energy assessment of large-scale hydrogen production via electrolysis and solar PV. Energy and Environ Sci. https://doi.org/10.1039/D1EE01288F
54. Parente CET, Lino AS, Carvalho GO, et al (2021) First year after the Brumadinho tailings' dam collapse: Spatial and seasonal variation of trace elements in sediments, fishes and macrophytes from the Paraopeba River, Brazil. Environ Res 193: 110526
55. Rehbein JA, Watson JEM, Lane JL et al (2020) Renewable energy development threatens many globally important biodiversity areas. Glob Change Biol 26:3040–3051
56. Rodríguez F, Moraga C, Castillo J et al (2020) Submarine tailings in Chile—A Review. Metals 11:780
57. Santamarina JC, Torres-Cruz LA, Bachus RC, RC, (2019) Why coal ash and tailings dam disasters ccur. Science 364(6440):526–528. https://doi.org/10.1126/science.aax1927
58. Schneider T, Kaul CM, Pressel KG (2020) Solar geoengineering may not prevent strong warming from direct effects of CO_2 on stratocumulus cloud cover. PNAS 117(48):30179–30185
59. Schramski JR, Gattie DK, Brown JH (2015) Human domination of the biosphere: Rapid discharge of the earth-space battery foretells the future of humankind. PNAS 112(31):9511–9517
60. Seibert MK, Rees WE (2021) Through the eye of a needle: an eco-heterodox perspective on the renewable energy transition. Energies 14: 4508 https://doi.org/10.3390/en14154508
61. Shugar DH, Jacquemart M, Shean D et al (2021) A massive rock and ice avalanche caused the 2021 disaster at Chamoli. Indian Himalaya. Science 373(6552):300–306. https://doi.org/10.1126/science.abh4455
62. Smallwood KS, Bell DA (2020) Effects of wind turbine curtailment on bird and bat fatalities. J Wildlife Mgt 84(4):685–696. https://doi.org/10.1002/jwmg.21844
63. Soltani M, Kashkooli FM, Souri M, et al (2021) Environmental, economic, and social impacts of geothermal energy systems. Renew Sustain Energy Rev 140: 110750
64. Staffell I, Green R (2014) How does wind farm performance decline with age? Renew Energy 66:775–786
65. Stenzel F, Greve P, Lucht W, et al (2021) Irrigation of biomass plantations may globally increase water stress more than climate change. Nature Comm 12: 512. https://doi.org/10.1038/s41467-021-21640-3
66. Sterman JD, Siegel L, Rooney-Varga JN (2018) Does replacing coal with wood lower CO2 emissions? Dynamic lifecycle analysis of wood bioenergy. Environ Res Lett 13: 015007
67. Stickler CM, Coe MT, Costa MH et al (2013) Dependence of hydropower energy generation on forests in the Amazon Basin at local and regional scales. PNAS 110:9601–9606
68. United Nations (UN) (2019) World Population Prospects 2019. https://population.un.org/wpp/
69. Van den Bergh J, Folke C, Polasky S et al (2015) What if solar energy becomes really cheap? A thought experiment on environmental problem shifting. Curr Opin Environ Sustain 14:170–179
70. Van der Zwaan B, Dalla Longa F (2019) Integrated assessment projections for global geothermal energy use. Geotherm 82:203–211
71. Van Haren H (2018) The pull of the tide. New Sci 23:24–25
72. Voigt CC, Straka TM, Fritze M (2019) Producing wind energy at the cost of biodiversity: a stakeholder view on a green-green dilemma. J Renew Sustain Energy 11: 063303. https://doi.org/10.1063/1.5118784

73. Weißbach D, Ruprecht G, Huke A et al (2013) Energy intensities, EROIs (energy returned on invested), and energy payback times of electricity generating power plants. Energy 52:210–221
74. Wikipedia (2021) Cost–benefit analysis. https://en.wikipedia.org/wiki/Cost-benefit_analysis
75. Williams JM (2020) The hydropower myth. Environ Sci Pollut Res 27:12882–12888
76. Wohlfahrt G, Tomelleri E, Hammerle A (2021) The albedo–climate penalty of hydropower reservoirs. Nat Energy 6:372–377
77. Zarf C, Berlekamp J, He F, et al (2019) Future large hydropower dams impact global freshwater megafauna. Sci Reports 9: 18531. https://doi.org/10.1038/s41598-019-54980-8

Chapter 4
Future Energy

Abstract There has never been a shortage of experts on the future. In this chapter, we first review the various methods in use for forecasting. Modern forecasting methods were not possible until reliable statistical data was available, and the simplest method merely extrapolates from past trends. The most popular approaches for energy forecasting make use of two methods, which are increasingly used in conjunction: scenario analysis, which looks at a number of plausible future paths, and integrated assessment models, which generate internally consistent models for future paths, given assumptions about future economic and technology developments. We critically assess the various forecasting methods in use (Sect. 4.2) before examining recent energy forecasts from both official energy organisations and major energy corporations (Sect. 4.3). In Sect. 4.4, we assess the impact of two important factors affecting energy forecasting: uncertainty in both EROIg calculations and climate sensitivity. Our conclusion is that, although we must engage in energy forecasting, trying to predict energy futures, never easy, is becoming increasingly difficult.

Keywords Back-casting · Deep uncertainty · Energy forecasts · Forecasting critiques · Forecasting methods · Integrated Assessment Models (IAMs) · Scenarios

4.1 Introduction to Energy Forecasting

In a Christian radio network broadcast a decade ago, the late Harold Camping, a retired civil engineer—like the first author of this book—announced that, based on his calculations, the world would end on 21 October 2011. His prediction—which was not his first for Judgement Day or the end of the world, stretching back as they did to 6 September 1994—was roundly criticised by atheist and Christian organisations alike [53]. Camping's prediction was precise, and certainly was 'important if true', with obvious energy policy relevance (don't bother building new energy infrastructure). Fortunately, for (most of) us, it was also incorrect. Camping's forecasts are just one of a vast series of apocalyptic predictions stretching back over many centuries, including the so-called Mayan prophecy forecasting cataclysmic change to the world a year later, on 21 December 2012. Once again, a no-show. Even supposedly science-based forecasts can get it badly wrong: the world is still waiting for flying cars, undersea cities, and colonies on the moon. On the other hand, the speed of information technology (IT) advances was greater than even many technological optimists had hoped for. As we will see, there is more to a prediction than simply getting it right, otherwise it is just a guess.

In summary, the world has always had its forecasters, claiming their ability to foretell the future. These acclaimed experts on the future include not only Harold Camping, but the Delphi oracles, Nostradamus, St Malachy, Mother Shipton, and other soothsayers, as well those basing their predictions on tarot cards and astrology. Nor has the growth of science over recent centuries done much to stem the popularity of these often-ancient methods. Entering the phrase 'predictions for the future' into the Google search engine returned over 14 million hits in mid-2021. Predictions are popular because most people want to know what the future holds, as even our actions in the present bear the imprint of how we view the future world.

Nor was there any shortage of popular and expert predictions made in 2019 for the year 2020, but very few predicted the key event—the pandemic that spread throughout the entire world. Nevertheless, the current pandemic did not take everyone by surprise. As Vaclav Smil [41] presciently reported in 2005: 'Expert consensus may be wrong, but it is certainly disconcerting to see that epidemiologists and virologists are in general agreement about a very high probability of pandemic influenza in the not too distant future—if not in a matter of months or years, then likely within the next one or two decades.' The experts couldn't give a precise year or even decade, but they did try to warn us that we should prepare for a major pandemic. The record shows that we didn't bother [25].

Despite the numerous failures for general forecasting, long-term energy fore-casting, albeit using far more sophisticated methods, nevertheless enjoys wide popu-larity. Noted energy researcher Vaclav Smil [42] is scornful of such attempts, arguing that 'Long-range energy forecasts are no more than fairy tales.' So, why are such energy forecasts still popular, and considered important? Possible answers include the need for energy-exporting countries to know the likely size of their markets in the importing countries, particularly if they are to develop new energy industries based on hydrogen, for example. Likewise, energy importing nations must worry about energy security and the availability of energy imports in the quantity required. And all countries must decide whether and how fast to expand—or even shut down—energy infrastructure, be it oil refineries, power stations, or electric power transmis-sion capacity [32]. Even if we usually get it wrong, we have no choice but to try to forecast. But none of this implies we have to try to develop forecasts as far ahead as the year 2100. The late futurist Graham Molitor was even more ambitious; his 2005 paper [29] included projections out to the year 3000. The last of his five 'big eras of economic activity', a new space age, he saw as 'commencing around 2500–3000'.

4.2 Energy Forecasting Methods—and Their Shortcomings

A wide array of forecasting techniques is available in the literature. At the same time, in addition to Vaclav Smil, there has been no shortage of strong critics of energy forecasting, or suggestions on how forecasts might be improved. Whether for global, national, or regional forecasts, all modern energy forecasts rely on accurate data about past and present energy use for the different types of energy. This is clearly the case when the simplest form of forecasting is used—extrapolating past trends into the future. A useful rule of thumb is to look back into the past as far as you are forecasting the future. Every decade has some more or less unexpected events important for energy use, such as the oil price shocks of the 1970s, or economic recessions which cut energy use. If the underlying trend was little affected by these events, it gives more confidence when extrapolating from the past to the future. Further, if the variations are cyclical, then the data must go back far enough to capture several such cycles, for example, the El Niño-Sothern Oscillation cycle in the Pacific Ocean. This method can be especially useful for short-term energy forecasts, but even forecasting energy use one year ahead can be difficult—as happened in 2020 with the economic downturn caused by the coronavirus pandemic. They can only work if the underlying conditions do not significantly change.

The recent history of air passenger transport illustrates the new fragility of forecasting. Although global energy use in general dropped in 2020, the greatest reduction occurred for aviation turbine fuel, as international passenger travel was severely curtailed. The aviation industry now thinks that full recovery for air travel cannot be expected until 2024, but given the rise of infection, even among the fully vaccinated [23], it is difficult to see how even vaccination certificates can enable air travel to completely return to normal. It is likely that tele-conferencing, which has now become an established feature, will continue to replace many face-to-face meetings and conferences. For tertiary education, online classes, introduced of necessity during lockdown, will probably still be used to some extent after the pandemic is under control—or, at least, at a level we can live with. This use of IT as a partial travel substitute will save fuel and emissions, as well as lower air pollution and travel costs. Transport, especially air travel, may have been changed forever.

The following subsections first discuss a variety of forecasting methods that have been used for general as well as energy forecasting (Sect. 4.2.1). Section 4.2.2 then critically examines Integrated Assessment Models (IAMs), mathematical modelling approaches which are heavily employed in energy and climate change forecasting.

4.2.1 General Forecasting Methods

Theodore Modis [28] has used the very observed close fit of the logistic curve to global energy use data from 1860 to 2017 to project energy use in 2050. His projected 2050 value was 25% higher than in 2017, and he boldly claimed a 90% chance of the value falling in the range 639–758 EJ. But sometimes, as the rapid spread of the delta variant of the pandemic virus is teaching us, the past can be a poor guide to even the very near future. In the Modis energy projection, no account was taken of any possible constraints on global energy use.

Also using the logistic curve, Cesare Marchetti [27] titled his ambitious 2009 essay '*On energy systems historically and in the next centuries*'. His claim is that not only energy sources but also transport modes have displaced earlier forms in a regular manner, with predictable intervals. So, coal replaced wood as the dominant form of energy, to be in turn overtaken as the leader in market share by natural gas. Similarly, motorised public transport replaced non-motorised travel, and cars replaced rail. The earlier forms did not disappear, but merely lost market share. Why have these changes occurred every few decades? One answer is technical progress— developments such as the steam engine or electric power—which allows new (and possibly cheaper and more abundant) energy sources to be used. It is not hard to see why fossil fuels replaced wood: energy use was no longer constrained by local annual biomass growth rates. Fossil fuels are a stock, so their use is only limited by the rate of extraction, not by annual biological limits—at least as long as reserves last. Furthermore, FFs have higher energy density, and coal, oil, and gas were readily transportable by ships and/or pipelines, allowing for large-scale imports in distant regions.

But what of the future? For energy, Marchetti saw FFs being replaced in the early decades of the twenty-first century by what he vaguely termed 'SOL-FUS'—solar and fusion energy. As we showed in Sect. 2.2, fusion energy production is most unlikely to be important for many decades—if ever. That leaves the various forms of solar energy. We agree with Marchetti's vague solar forecast—most RE sources are ultimately solar—but what it implies is that instead of ever-novel forms of energy, we are reverting to older forms, albeit with new technologies for energy conversion. The finite nature of FF and even nuclear fuel reserves, and their environmental burden, will eventually see to that. Although not discussed, the move back to solar will, presumably, be the last fuel cycle.

Another approach—the Delphi Method—a nod to the famous oracle of Delphi, once consulted in the ancient world for all important decisions—involves canvassing the opinions of experts on a given topic. It's an iterative process, in which the answers given by any respondent are shared with the others, and respondents can modify their responses in light of this feedback. The question of interest might be, for example: what share of the market will hydrogen energy have in various energy sectors by a given year? Valette et al. [48] in 1974/5 polled 86 experts on this question of future developments in hydrogen (H2) use. Their median forecasts for H2 use in transport greatly underestimated the time for hydrogen to achieve significant market share. They foresaw H2 market share in the year 2000 as 10% for both private road and air travel, and 20% for buses. The actual values were zero or near zero for all modes. Further, liquid H2 rather than compressed H2 was (wrongly) seen as the favoured onboard storage technology; today, it has been largely abandoned for road transport in favour of compressed H2.

Is the Delphi method any improvement on other methods? Some researchers don't think so. The method has long been subject to doubts concerning its accuracy and possibilities for bias, with one study concluding that experts in a given subject area perform no better than generalists [30]. Looking at forecasting more generally, a study of forecasting political events made in 2005 found that experts in a given speciality performed far worse than expected by chance [36]. Experts also perform poorly on the economic future: most expert forecasts for the economy, a topic of great interest and importance, did not foresee the 2008 Global Financial Crisis [26]. Makridakis et al. [26] also quote Steve Ballmer, the CEO of Microsoft, as commenting in 2007: 'There's no chance that the iPhone is going to get any significant market share'.

However, this may be too dismissive of more recent efforts to refine the method. Kattirtzi and Winskel [17] argue a variant of this method, Policy Delphi, can help decision-makers better understand the disagreements among experts, and their causes, including diverging values systems. In any case, as discussed above by Smil [41], we should have listened more carefully to what disease experts were saying in the years up to the Covid-19 outbreak. Sometimes, we need experts who are fixated on one topic, and who have important insights for the likely future.

Zellner and colleagues [55] reviewed the relative merits of human judgement versus quantitative methods for forecasting. With human judgement approaches, forecasters posit futures based on their plausibility, whereas quantitative forecasting attempts to calculate the probabilities of various outcomes. They found a very large rise in the popularity of quantitative forecasts, but concluded that neither approach is clearly superior; it depends on such factors as data availability—and, of course, the topic being investigated. Often, a combination of both approaches will give the best outcomes, the conclusion also reached by Scoblic and Tetlock [36].

What do we mean by 'better' forecasts, anyway? For energy, forecasting might mean predicting total primary energy use at the national or global level. Or, energy costs could be the focus, as in the many studies trying to forecast international petroleum prices. Or, the interest could focus mainly on just one energy source, as is the case for specialist energy organisations serving the petroleum, nuclear, wind, solar, and other energy type industries. It is also possible to arrive at a correct prediction, but still provide misleading answers for policymakers. Smil [40] recounted how his 1975 projections for China's commercial energy use in 1985 and 1990 turned out to be quite accurate. However, the accuracy was quite fortuitous: he not only underestimated China's economic growth by a factor of two, but also underestimated the decline in energy intensity by about the same factor!

The *consequences* of inaccurate forecasts will vary depending on circumstances. If 2020 electricity forecasts were overestimated in most OECD countries, it didn't really matter (except to the electricity supply industry), since all it meant was that there was spare capacity available. But if an exceptionally cold winter means that heating energy demand is underestimated, it can lead to fuel shortages and social unrest. Nevertheless, the accuracy of any forecast may not always be a good indicator of its utility. The existence of *self-fulfilling* and *self-defeating* (or *self-negating*) predictions has been recognised for many decades [35]. The classic example of a self-fulfilling prophecy is that rumours of a bank collapse can lead to mass withdrawals of deposits, making collapse more likely. Conversely, self-defeating predictions cause the prediction to fail, as a clear warning of imminent danger can often lead to policy changes being made to avert the crisis. But such warnings do not always have this effect, as the ignored warnings about the current pandemic, and decades of increasingly more urgent IPCC warnings about climate change, make clear. One reason might be that such dire warnings might alienate the public, if a feasible path to avoid the danger is not perceived as available [9]. Another reason may be that we have no living connection to similar past events.

The important point here is that predictions themselves can influence the future they are trying to predict—they are one of the inputs into how planners and the general public *imagine* the future. A transport policy example from the UK is the 'predict and provide' approach to road building, where official forecasts of rising traffic justified road construction—which did lead to increased traffic [11]. But given the plethora of future energy forecasts discussed in the previous section, it is doubtful that any one of them can have a policy impact similar to the single official UK traffic forecast. Yet one assumption is consistent across all the scenarios, with their huge range of numerical energy forecasts: that global economic growth will continue indefinitely.

4.2.2   Integrated Assessment Models (IAMs)

Integrated assessment models (IAMs) are very popular in formulating future energy forecasts, and have been extensively used in successive IPCC forecasts to assess future FF energy use and the resulting GHG emissions. Criticisms of IAMs now go back nearly several decades; more recently, their use has been criticised by climate scientist Kevin Anderson [3], who stated in a *Nature* article that they are simply the 'wrong tool for the job'. In a companion piece response, Jessica Jewell [16] defended their use. She pointed out that the first use of IAMs in the climate debate was to generate scenarios showing the path GHG emissions would take under an internally consistent set of technological and economic changes, with plausibility an important component for scenario selection. She also credits IAMs with showing how continuing on a business-as-usual path would result in dangerous CC later in this century.

Spangenberg [43, 44], among others, has helped put IAMs in the more general context of complexity theory. He points out that 'a "system" is any set of things within a common frame (the system boundary) that is ruled by a given set of interactions (the system rules)'. Overall, three systems need to be distinguished:

- The real-world system, or in plain terms, reality.
- Our mental models of that reality, necessarily a simplified model.
- Computer models, 'the tools used to quantify a selected set of the expectations raised by the mental models'.

While system boundaries can be arbitrarily defined, the same is not true of the system rules which define the functioning of the system. The authors list five rules to describe system complexity, as follows:

1. The system can be distinguished from its environment.
2. System components can be distinguished, and their interactions described.
3. The active system elements (such as CO_2 molecules, or, in economics, consumers or producers) behave identically.
4. 'The individual behaviour of the system elements can be described by average interaction parameters which characterise the system behaviour'.
5. 'The system develops towards a stationary equilibrium', for example, zinc reacting with hydrochloric acid. The system and its end products are then perfectly predictable.

The movement of the planets meets all five rules, but Earth's climate only satisfies rules 1–4, in that it can adapt, and have tipping points. Biological systems only meet rules 1–3, as individual agents do not necessarily behave in a set way. Evolution can occur. With humans in the system, only rules 1, and perhaps 2, are met. As we saw with self-negating prophecies, humans can change their behaviour in the face of adverse consequences for standard behaviour. As the authors argued: 'taking this trait into account is a necessary condition for suitable economic models [...].' It follows that we must be extremely careful in interpreting the output of IAMS, particularly since the simplifying assumptions underpinning the model are not usually evident. Instead, users tend to focus on the numerical and graphical outputs. The data—often given to several significant figures—can instill a false sense of certainty in the future.

Another serious shortcoming of IAMs—in fact most energy forecasts—is that the *equity* dimension is almost totally ignored. Chapter 1 stressed the orders of magnitude differences in average per capita electricity use between countries. Yet, as we saw in Chap. 3, proposed major hydro developments, mostly in low- or middle-income countries, and vastly increased bioenergy output, as in the BECCS proposals in IPCC [12], will often disadvantage already marginal groups. The benefits from large hydro projects can be large, but are reaped largely by cities and industry. For local people, however, the costs can be high. Large dams have led to the displacement of an estimated 80 million people over the past century, often without adequate compensation, and loss of livelihood for them and others living downstream of the project [21]. Even if net benefits do outweigh the costs, the winners seldom provide adequate compensation for the project losers. As the authors argued: 'taking this trait into account is a necessary condition for suitable economic models [...]'.

A final criticism of IAMs concerns the *costing* of the various energy sources. As discussed in Sect. 3.2, the full energy costs of RE projects are often underestimated, just as they are for FFs. These ecosystem maintenance energy (ESME) costs will also incur a monetary cost, which means that models are usually working with inaccurate cost data from a sustainability viewpoint, further limiting their use in energy or climate change policy. Steve Keen [19] in an article entitled *'The appallingly bad neoclassical economics of climate change'* is extremely critical of estimates of economic damages from climate change produced by assessment models such as DICE. He pinpoints three doubtful assumptions/procedures made by economists:

- Assuming that 'about 90% of GDP will be unaffected by climate change, because it happens indoors'.
- Use of present temperature-GDP relationship between countries (and US states) to gauge the impact of global warming over time.
- Reliance on the optimistic climate-related views of economists, rather than the better-informed warnings from scientists.

He concluded that 'Correcting for these errors makes it feasible that the economic damages from climate change are at least an order of magnitude worse than forecast by economists, and may be so great as to threaten the survival of human civilization.'

4.3 Global Forecasts from Energy Organisations

Global energy forecasts out to years 2040, 2050, or even 2100 are available from various national and global energy organisations such as the US Energy Information Administration (EIA), the International Energy Agency (IEA), the Intergovernmental Panel on Climate Change (IPCC), the Organization of the Petroleum Exporting Countries (OPEC), and the World Energy Council. Additionally, energy-related corporations—BP, ExxonMobil, and Shell—also regularly publish global forecasts. These forecasts we have discussed in detail in [33], where the large spread of values, both for 2050 projected primary energy use, and for the RE share of this total, is stressed.

One important reason for this spread of values is the rising use of energy *scenarios (or pathways)*. In earlier decades, forecasts were often single-value, with perhaps a sensitivity analysis giving minor variations from the central forecast. The annual EIA forecasts of the US still use this approach [8]. But increasingly, scenario analysis is employed, as in the BP and IEA forecasts [4, 14]. These often take the form of examining the effects of different climate change mitigation policies on energy use and type. The BP forecasts, for instance, use three scenarios for the year 2050, which they term business-as-usual (BAU), Rapid, and Net Zero (for net CO_2 emissions in the year 2050). All scenarios assume further global energy growth from 2018 to 2050, with BAU growth the highest. For net CO_2 emissions in 2050, BAU, Rapid, and Net Zero scenarios assume 30.5, 9.4, and 1.4 Gt/year, respectively, compared with 34.0 Gt of CO_2 energy-related emissions in 2019.

The forecasts of the oil company Shell [38] are interesting for two reasons. First, Shell pioneered the use of scenarios as an aid to decision-making. Second, Shell's recent scenarios were devised with the current pandemic in mind. Shell presented three scenarios, with the unhelpful titles 'Waves', 'Islands', and 'Sky 1.5', each representing different speeds of decarbonisation: 'late but fast'; 'late and slow'; and 'accelerated now', respectively. Corresponding predicted global temperatures above pre-industrial in the year 2100 are 2.3 °C, 2.5 °C, and 1.5 °C, respectively.

All scenarios give values of primary energy use by type, out to the year 2100, as well as global GDP (in constant $US 2016 purchase parity pricing (PPP) values). Energy use is also broken down by both economy sector and geographical region. All three scenarios show large increases in both total primary energy use and RE use, with corresponding declines in all FFs. In all three, real global GDP also rises several-fold out to 2100, with the Waves scenario having both greatest GDP and RE uptake. What conclusions can we draw from these forecasts? The main one is that the further we look into the future, the greater is the range of possibilities envisaged; by the year 2100, most things look possible. This contradicts ecological economists like Daly [6] and ecological footprint researchers, who think that, already, the disbenefits of growth exceed the benefits, so that further economic growth is problematic.

Other forecasts might be implicitly or explicitly *normative*, as in back-casting, where a desirable future is postulated, and the changes in various parameters compatible with this end-state are explored. 'Desirable' forecasts beg the question: desirable for whom? FF energy companies (and even FF exporting nations) will answer the question differently from advocates of a RE energy transition. A good example of such normative forecasting are the four scenarios of the IPCC [12], used to explore the options for meeting the 1.5 °C limit on global temperature rise.

Scenario forecasting has clear limits for many applications. Whether planning a new airport, or a factory expansion to expand product output, it is of little help for the relevant decision-makers to be told that annual airport passengers, or product sales, could vary by a factor of 10, for example. Futurists often categorise futures into possible, preferred, and probable futures. Preferred (or normative), as well as probable futures, evidently must be selected from the set of possible futures. But how probable are the scenarios of the IPCC [12], with their very high levels of biomass energy use, combined with BECCS? Anderson and Peters [2] are sceptical of BECCS and other NETs, labelling them as unproven (see also the discussion in Sects. 2.4 and 3.4.3). Rogelj et al. [34] have discussed scenarios with very high levels of RE overall in the year 2100 (1000 EJ) including 400 EJ from BECCS. From our discussion in Chapter 3, these high energy figures lack plausibility.

As a 'what if' scenario, the inclusion of some BECCS in the IPCC and other scenario sets is perhaps reasonable; what is not reasonable is the refusal to even countenance the end of economic growth as a possibility—whether planned or unplanned—a topic we shall treat more fully in Chap. 5. Also, any energy reductions are seen as the result of unprecedented large drops in *energy intensity*—the energy needed per unit of GDP, not from energy conservation. High bioenergy and BECCS may be possible—but in our view are less probable than many possible scenarios that have been ignored.

As an illustration, Fig. 4.1 shows projections for various years for the global installed capacity of nuclear energy. It can be seen that the early estimates for nuclear power were far too optimistic. The nuclear forecasts for the year 2000 made in the early 1970s illustrate the point that given enough time, very large increases appear possible, even probable. But as the forecast horizon shrinks, forecasts become more realistic. Because of the long lead times for nuclear plant planning, approval, and construction, nuclear energy output forecasting is now easier than for other energy

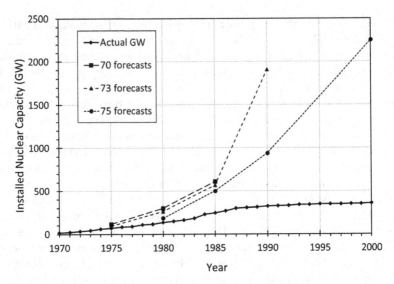

Fig. 4.1 Some energy forecasts for nuclear power versus year of forecast. Note that the forecasts are for 'World outside the Communist countries', so global forecasts would be even higher. The 'Actual GW' curve is for global installed capacity. *Source* [22]

sources, at least out to two decades. In another area, forecasts in recent decades for the global population have tended to be accurate, chiefly because of its strong dependence on future population on the existing levels. Before that time, errors were large, because earlier statistics were in error [31].

Dynamic energy analysis offers a further insight as to why the rapid deployment of RE discussed in forecasts such as the 'Net Zero' scenario of BP and Shell's 'Sky 1.5' are unlikely [5, 20]. In Chap. 3, we showed that when a full EROI accounting is done to include ESME costs and the need for energy storage as intermittent energy sources start to dominate total energy supply, EROI values may be quite low, even if still greater than unity. Further, economies need an EROI to be much greater than unity for a good quality of life [24]. What happens if we try to rapidly raise the output from RE sources, in an effort to respond quickly to the need for CC mitigation? More and more output energy will need to be diverted as input energy into constructing new RE energy systems. The net energy output available to the non-energy sector of the economy will consequently fall. The only responses would be to use less energy, or to increase FF output, once again putting CC mitigation in jeopardy. Or, another possibility, RE providers could continue to largely ignore ESME costs as is the present practice, but this will only lead to their deferral—or perhaps to irreversible ecological damage, which will likely negatively affect future RE production later on.

Predicting the future *costs* of energy sources is even more difficult than forecasting the quantities used. Nowhere is this more evident than for international oil prices. In real terms, oil prices have varied by as much as an order of magnitude over the past half-century—or by even more if the briefly *negative* prices in 2020 are included. The causes for price fluctuations are varied, and include organised oil supply cutbacks by oil-exporting countries, as well as economic cycle variations in demand, the presence of cold northern hemisphere winters [15, 18], or even by unforeseen events such as blockage of the Suez Canal by single container ship, for 6 days as occurred in March 2021. Large price rises can damage importing country economies; conversely, large price falls damage those of oil-exporting countries.

4.4   Multiple Sources of Future Uncertainty

To fully appreciate how tenuous our grasp on future energy use really is, we need to examine the several factors feeding into this uncertainty. These include uncertainty in our knowledge concerning.

- ESME costs and EROI for various RE sources, and how these will change over time.
- Crucial biophysical parameters such as climate sensitivity and ecosystem response to changing climate.
- The likely policy response of decision-makers to climate change and other ecological threats.
- The extent to which global inequity in wealth is tackled.

The first two factors will be examined in this chapter, the remaining two in Chap. 5. In Chap. 3, we discussed not only the need for EROIg analysis for energy decision-making but also uncertainties surrounding its calculation. Much of the uncertainty hinges on the calculation and inclusion of the ESME costs, particularly those for mining the minerals needed for the construction of the energy conversion devices. Without a good EROIg analysis, we cannot say whether a new energy project is viable from an ecological sustainability viewpoint. This, of course, has not stopped the many FF plants still under construction or approved going ahead. The ESME costs of RE are smaller than those for FF, but far from negligible, and can even overlap with those for FF per unit of net output energy.

Their uncertainty means that not only can we not accurately determine EROIg values but we have also no real idea of the annual level of net green RE energy potentially available to the world, at or above a given EROIg cut-off point. It may well be that the total is well below our present level of global energy use [33]. It's even quite possible that increasingly in future, we will initiate projects that give less than unity values for EROIg—as long as they can deliver net monetary returns. Some researchers think that one set of projects—palm oil plantations for biodiesel in Malaysia and Indonesia—already fall into this category. The possible implications of this will be explored in Chap. 5. Some hydropower schemes can also deliver negative returns, as discussed in Sect. 3.4.4. Energy systems that fail to operate over their expected life due to changed circumstances can be included in this category.

Uncertainty also is a feature of the models used to simulate climate change. Mathematical models are the mainstay of the 'hard sciences', particularly physics, where they have proved remarkably successful. They can make reliable absolute predictions, such as when Halley's comet will make its next appearance (see Sect. 4.2.2). Predictions of this type are possible because the physics of cometary paths is well understood, and human actions cannot (at least at present) influence their paths. In climate science, mathematical models are extensively used for *conditional* predictions, as in answering the important question: If the concentration of equivalent CO_2 in the atmosphere doubles, what will be the equilibrium rise in global surface temperature? Even for this conditional prediction, no definite answer has been forthcoming. Values for this parameter, termed the *climate sensitivity* of the Earth system, were, until recently, little more accurate than the answer given by the Swedish scientist Svante Arrhenius a century ago.

Not only is the actual value of equilibrium climate sensitivity still unknown, but we have little idea of its possible range, particularly at the upper end. In a long review in the year 2021, Sherwood and colleagues [39] attempted to put tighter bounds on estimates of climate sensitivity. They concluded that a lower limit was 2.0 °C, as lesser values were difficult to reconcile with the combined evidence from several approaches. The upper limit was less closely bounded, but could reach as high as 5.7 °C. A disagreement was found between the modelled results and those that place more reliance on the actual path global temperature has taken in recent decades. The modelled results gave some implausible values, and so were adjusted to better fit observation [50]. A *New Scientist* article from mid-2021 simply posed the question: 'Is the climate becoming too extreme to predict?' [49].

The fifth IPCC report stated that there is a 66% probability that it lies between 1.5 and 4.5 °C, which would imply a one in three chance of being outside that range. The sixth IPCC report halved the range to 2.5 and 4.0 °C at the same probability level [13]. But, as Nassim Taleb [46] has stressed for some time, the real worry lies with what he terms 'black swan' events. By this term, he means the low probability but very high-risk events—such as the risk of climate catastrophe. It is all very well for IPCC [13] to imply that we only have a 33% chance of the climate sensitivity value exceeding 4.0 °C. But would any of us want to travel on an airline that advertised that the chance of a crash was 'only' 33%? And compared to other existential risks, such as for a large asteroid impacting earth, 33% or even 10% is a very high value. We need to adopt the precautionary principle, given the high risk of irreversible damage. Past a certain point, accuracy becomes unimportant.

4.5   Concluding Comments on Energy Forecasting

As we have shown, it is very easy to criticise forecasting in general, and energy-economic forecasts in the form of IAMs in particular. But we have to try to get some sort of handle on the future: we have no choice. The best approach for doing this will vary according to what we are trying to forecast. It is also important to realise, not only that some forecasts are easier to make than others but also that the *consequences* of wrong forecasts matter far more in some areas than others. With much simplification,

Fig. 4.2 Classification of forecasts by reliability and importance of consequences

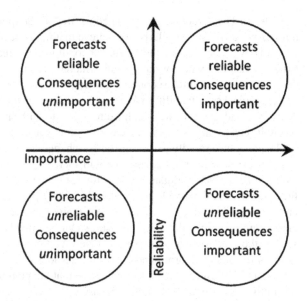

forecasts can thus be conveniently placed into one of four quadrants, as shown in Fig. 4.2. The year 2000 forecasts for nuclear power output made in the mid-1970s were an example of the lower left quadrant type. They proved to be wildly in error, but the consequences were minor, for two reasons. First, later forecasts were far more modest. Second, nuclear is only one of a number of methods for generating electricity, so that if one source produces less than expected, others can take up the slack.

For energy forecasting, we can learn much from the experience of modellers trying to predict the path of the pandemic at national and global levels. One very obvious difference is that none of these pandemic predictions tried to look very far ahead, certainly not to the year 2050 or 2100. Instead, they merely attempted to predict a few months to, at most, a few years ahead, and even that proved very difficult [37], as subsequent events (especially the rise of the delta and omicron variants) demonstrated. The trajectory graphs of key parameters, such as daily mortality or intensive care unit beds needed for Covid-19 patients per million population developed by Walker et al. [51], only attempted a forecast over a period of 100–500 days. Earlier, we mentioned the need for underlying conditions to remain constant if extrapolation is used for prediction. This approach will not work for predicting the path the pandemic will take. The reason? The virus can evolve very rapidly, producing new variants with different health-related risks [10]. Further complicating the job of modellers is the response of governments and the public. Mask-wearing, social distancing and lockdowns, and the extent to which they are followed are difficult (or impossible?) to model accurately.

In a detailed study for the US, Weiskopf et al. [52] reported that climate change would produce mixed effects on biodiversity and ecosystem services, but that even the positive effects would need 'costly societal adjustments' for these benefits to be realised. The authors urged the need for flexible adaptation strategies in order to minimise these costs. Similarly, we believe that many decisions about future energy use will have to be made on the run, under conditions of great uncertainty. We do need to have a vision of where we are going, which we argue must be one that replaces the dominant one of continued economic growth. There are two parts to this vision: equity and ecological sustainability. We need to abandon dreams of forecasting energy use decades ahead. Instead, we will have to make do with short-term forecasts in areas where it really matters—as is the case with the pandemic—and will need to be able to change course rapidly as circumstances change.

As average temperatures rise, we will increasingly need climate *adaptation* as well as do our best to mitigate further climate changes. Adaptation, unlike mitigation, faces few political obstacles; it is already widely practiced—think of fire brigades and emergency services—and the benefits mainly accrue to those countries or groups employing it. Some adaptation measures can be quickly implemented, but others, such as the proposed barriers to protect New York from storm surges and rising seas in the wake of superstorm Sandy, can take decades to build and cost tens of billions of dollars [47]. But we need to make sure that adaptation measures do not hinder CC mitigation efforts. One important case is the possible greatly increased purchase and use of air conditioners as average temperatures rise globally. They may even be needed for continued habitability in parts of the world, as we discussed in Sect. 1.4. Unless the world moves quickly to RE electricity, their use will only exacerbate global warming—and the need for even more air-conditioning.

This need for flexibility will also favour some forms of low-carbon energy over others. The modular nature of wind and PV energy farms, with their short construction times, will be favoured over hydropower (and nuclear) plants. Even their relatively short service lives (20–30 years) need not be a problem, since it is possible that ongoing climate change will in many cases alter output drastically over longer time frames. The high EROI values usually claimed for hydro plants are mainly the result of service lives of up to a century, but stable output for a century is unlikely to be realised for new plants, if climate continues to worsen.

In light of the above warnings, what can be usefully said about global energy futures? It is probable that energy consumption (especially FF use) in the core OECD countries will continue to fall; this merely continues a trend that is decades long, and partly addresses the need for CC mitigation. Outside the OECD, energy forecasting is a perilous task. FF use is still rising along with RE use. All we can usefully give is a normative forecast; overall energy consumption will need to rise in presently low per capita energy use countries to satisfy basic needs, and this additional energy should be provided from local, genuinely green RE sources. In this endeavour, these countries will need technical and economic support from high-income nations. If core OECD nations do cut FF use significantly and rapidly, excess FF capacity will occur, cutting their need for rapid RE introduction there.

We started this chapter with examples of failed apocalyptic warnings about the future. Nevertheless, we ourselves need to be careful we don't fall into a similar trap, and embrace what has been called 'apocalyptic environmentalism' [45]. As Alexander et al. [1] stress, in attempting to forecast energy futures—or indeed, futures of any type—it is necessary to employ 'knowledge humility'. Major changes in both the biophysical world and the socio-economic world mean that forecasting, always difficult, will progressively become even more so. Given the increasingly dire warnings by climate and other scientists of the crises we face [7, 9], the problems discussed in Chap. 1 are unlikely to be overstated—and may even be understated. It is at least possible that one of the several solutions put forward for solving these crises might work, and technological progress will likely continue. But as discussed in Chap. 2, the risk of unwanted side effects, both foreseen and unforeseen, are likely to multiply as we push towards Earth's several limits.

References

1. Alexander S, Floyd J, Lenzen M et al (2020) Energy descent as a post-carbon transition scenario: How 'knowledge humility' reshapes energy futures for post-normal times. Futures 122:1025652
2. Anderson K, Peters G (2016) The trouble with negative emissions. Science 354:182–183
3. Anderson K (2019) Wrong tool for the job. Nature 573:348–349
4. BP (2020) BP Energy outlook 2020. London, BP
5. Capellán-Pérez I, de Castro C, González LJM (2019) Dynamic energy return on energy investment (EROI) and material requirements in scenarios of global transition to renewable energies. Energy Strategy Rev 26: 100399
6. Daly H (2019) Growthism: its ecological, economic and ethical limits. Real-World Econ Rev 87:9–22
7. Díaz S, Settele J, Brondízio ES, et al (2019) Pervasive human-driven decline of life on Earth points to the need for transformative change. Science 366, eaax3100
8. Energy Information Administration (EIA) (2019). International energy outlook 2019. https://www.eia.gov/outlooks/ieo/pdf/ieo2019.pdf. (Also, earlier editions).
9. Folke C, Polasky S, Rockström J, et al (2021) Our future in the Anthropocene biosphere. Ambio 50: 834–869 https://doi.org/10.1007/s13280-021-01544-8
10. Goodman J (2020) A genetic gambit. New Sci 17 July: 22
11. Goulden M, Ryley T, Dingwall R (2014) Beyond 'predict and provide': UK transport, the growth paradigm and climate change. Transp Pol 32:139–147
12. Intergovernmental Panel on Climate Change (IPCC) (2018) Global warming of 1.5 °C: summary for policymakers. Switzerland, IPCC (ISBN 978–92–9169–151–7)
13. Intergovernmental Panel on Climate Change (IPCC) (2021) Climate change 2021: the physical science basis. AR6, WG1. CUP, Cambridge UK (Also earlier reports)
14. International Energy Agency (IEA) (2020) Renewables 2020: Analysis and forecast to 2025. https://webstore.iea.org/download/direct/4234.
15. Jawadi F (2019) Understanding oil price dynamics and their effects over recent decades: an interview with James Hamilton. The Energy J 40(SI2):1–13
16. Jewell J (2019) Clarifying the job of IAMs. Nature 573:349
17. Kattirtzi M, Winskel M (2020) When experts disagree: using the Policy Delphi method to analyse divergent expert expectations and preferences on UK energy futures
18. Kaufmann RK, Connelly C (2020) Oil price regimes and their role in price diversions from market fundamentals. Nat Energy 5:141–149
19. Keen S (2020) The appallingly bad neoclassical economics of climate change. Globalizations. https://doi.org/10.1080/14747731.2020.1807856
20. King LC, van den Bergh JCJM (2018) Implications of net energy-return-on-investment for a low-carbon energy transition. Nat Energy 3:334–340

21. Kirchherr J, Ahrenshop M-P, Charles K (2019) Resettlement lies: suggestive evidence from 29 large dam projects. World Dev 114:208–219
22. Krymm R, Woite G (1976) Estimates of future demand for uranium and nuclear fuel cycle services. IAEA Bull 18: 5/6 (Also earlier and later IAEA bulletins)
23. Kupferschmidt K (2021) New SARS-CoV-2 variants have changed the pandemic. What will the virus do next? Science https://www.sciencemag.org/news/2021/08/new-sars-cov-2-variants-have-changed-pandemic-what-will-virus-do-next
24. Lambert JG, Hall CAS, Balogh S, Gupta A, Arnold M (2014) Energy, EROI and quality of life. Energy Policy 64: 153–167
25. Mackenzie D (2020) The covid-19 pandemic was predicted – here's how to stop the next one. New Sci 16 Sept https://www.newscientist.com/article/mg24733001-000-the-covid-19-pandemic-was-predicted-heres-how-to-stop-the-next-one/#ixzz71lqPvS2H
26. Makridakis S, Hyndman RJ, Petropoulos F (2020) Forecasting in social settings: the state of the art. Int J Forecasting 36:15–28
27. Marchetti C (2009) On energy systems historically and in the next centuries. Global Bioethics 22(1–4):53–65
28. Modis T (2019) Forecasting energy needs with logistics. Technol Forecast Soc Chang 139:135–143
29. Molitor GTT (2005) From my perspective: five economic activities likely to dominate the new millennium. VII: Principles and patterns of economic era development Technol Forecast Soc Chang 72: 85–99
30. Moriarty P, Honnery D (2011) Rise and fall of the carbon civilisation. Springer, London
31. Moriarty P, Honnery D (2018) Three futures: nightmare, diversion, vision. World Futures 74(2):51–67. https://doi.org/10.1080/02604027.2017.1357930
32. Moriarty P, Honnery D (2019) Energy accounting for a renewable energy future. Energies 12:4280
33. Moriarty P, Honnery D (2021) The limits of renewable energy. AIMS Energy 9(4):812–829. https://doi.org/10.3934/energy.2021037
34. Rogelj J, Popp A, Calvin KV et al (2018) Scenarios towards limiting global mean temperature increase below 1.5 °C. Nature Clim Change 8:325–332
35. Sabetta L (2019) Self-defeating prophecies: When sociology really matters. In: Poli R, Valerio M (eds) Anticipation, agency and complexity, anticipation science 4, Springer Nature Switzerland AG. https://doi.org/10.1007/978-3-030-03623-2_4
36. Scoblic JP, Tetlock (2020) A better crystal ball: the right way to think about the future. Foreign Aff 99 (6): 10–18
37. Scudellari M (2020) The pandemic's future. Nature 584:22–25
38. Shell International BV (2021) The energy transformation scenarios. www.shell.com/transform ationscenarios
39. Sherwood SC, Webb MJ, Annan JD, et al (2021) An assessment of earth's climate sensitivity using multiple lines of evidence. Rev Geophys 58: e2019RG000678. https://doi.org/10.1029/2019RG000678
40. Smil V (2000) Perils of long-range energy forecasting: Reflections on looking far ahead. Technol Forecast Soc Chang 65:251–264
41. Smil V (2005) The next 50 years: Fatal discontinuities. Pop Develop Rev 31(2):201–236
42. Smil V (2008) Long-range energy forecasts are no more than fairy tales. Nature 453(7192):154
43. Spangenberg JH (2018) Looking into the future: Finding suitable models and scenarios. In: Düwell M, Bos G, van Steenbergen N (eds) Ch 4 in Towards the ethics of a green future. Routledge, UK
44. Spangenberg JH, Polotzek L (2020) Challenges of complexity economics. WEA Commentaries 10(1):8–11
45. Swyngedouw E (2011) Depoliticized environments: the end of nature, climate change and the post-political condition. Roy Inst Philos Suppl 69:253–274. https://doi.org/10.1017/S13582 46111000300
46. Taleb NN (2013) Antifragile: Things that gain from disorder. Random House and Penguin

47. Tollefson J (2013) New York vs the sea. Nature 494:162–164
48. Valette P, Valette L, Siebker M et al (1978) Analysis of a Delphi study on hydrogen. Int J Hydrog Energy 3:251–259
49. Vaughan A (2021) Is the climate becoming too extreme to predict? New Sci 31 July: 11
50. Voosen P (2021) Climate panel confronts implausibly hot models. Science 373(6554):474–475
51. Walker PGT, Whittaker C, Watson OJ (2020) The impact of COVID-19 and strategies for mitigation and suppression in low- and middle-income countries. Science 369:413–422
52. Weiskopf SR, Rubenstein MA, Crozier LG, et al (2020) Climate change effects on biodiversity, ecosystems, ecosystem services, and natural resource management in the United States. Sci Total Environ 733:137782
53. Wikipedia (2021) List of predictions https://en.wikipedia.org/wiki/List_of_predictions
54. World Energy Council (WEC) (2019) World energy scenarios 2019. WEC, London
55. Zellner M, Abbas AE, Budescu DV, et al (2021) A survey of human judgement and quantitative forecasting methods. R. Soc Open Sci 8: 201187 https://doi.org/10.1098/rsos.201187

Chapter 5
Discussion and Conclusions

Abstract In this chapter, we begin with a summary of the key themes of this book: the serious ecological sustainability problems our planet faces; that in the Anthropocene, these problems are increasingly of our own making; that technological solutions are less effective because solutions to one problem can aggravate the other problems; that global inequality is high and still rising. The chapter then discusses some possible changes that could both improve ecological sustainability and global equity. The first of these changes would address the unpaid external costs of FF combustion and use, possibly through carbon taxes. Then the possible changes to transport and agriculture, both important energy-consuming sectors, are used as examples. We then assess the feasibility of these changes, arguing that the lessons from the current epidemic, together with the rise of extreme weather events experienced by an ever-increasing share of the global population, open up space for more rapid social and economic changes. Nevertheless, while the changes are possible to achieve, success is not guaranteed, especially in the near term.

Keywords Climate change · Degrowth · Energy equity · Energy limits · Future energy · Global pandemic · Renewable energy

5.1 Introduction: The Key Themes of This Book

This book has focused on several themes. First, that our planet is facing a number of very serious biophysical problems, as exemplified by the planetary boundaries concept, as elaborated in Chap. 1. We are approaching the safe limits for many of these boundaries, including biosphere integrity and ocean acidity. For climate stability, we may have already crossed the threshold, as evidenced by the alarming rise in extreme weather events throughout the world, as discussed in Sect. 1.2. The Ecological Footprint (EF) analysts have reached a similar conclusion, although using a different approach; leading EF proponent Mathis Wackernagel [54] argues that 'humanity is currently using nature 1.7 times faster than ecosystems replenish, akin to using 1.7 Earths.' Problem: We only have one Earth.

The second theme is that the dire environmental problems we face are increasingly of our own making. We have now entered the Anthropocene epoch [29]. The previous epoch, the Holocene, which lasted from the end of the last glaciation about 12 millennia ago until very recently, enjoyed a fairly stable climate, which enabled early civilisations to flourish. We are still partly at the mercy of the vagaries of climate, but increasingly, the changing climate we are already beginning to experience is the consequence of our own actions. As climate scientists would say, the signal-to-noise ratio has risen, and the signal has become abundantly clear. No longer can future climate be accurately predicted, even in principle, using only biophysical science methods. Social and political concepts are now also needed, and we must find ways of integrating these different approaches, with their very different methods of study.

The third theme is that, regardless of their past successes, technological solutions to our many and interrelated biophysical problems have increasingly been found wanting. Chapter 2 examined a number of these relevant to energy, including nuclear power, energy efficiency improvements, CDR, and SRM, while Chap. 3 looked at a key component of the increasingly popular 'green growth' solution—rapid substitution of FFs by RE sources. All were found to be defective in some way, when examined in the light of the full range of biophysical problems facing us. In many cases, the proposed solution to one major challenge, climate change, merely led to new risks in other areas. The risk of catastrophic climate change is, unfortunately, not the only crisis we face. In Sect. 1.2, we saw that many researchers think that biodiversity loss, with the attended losses in ecosystem services that result from our shrinking natural capital, is just as important—if not more so.

The solution suggested in this final chapter is deep and rapid energy conservation, by which we mean using fewer energy-consuming devices and/or using them less. In addition, the world will have to largely replace FFs with RE sources, but a simple rapid substitution of FFs by RE does not appear possible, for the following two reasons. First, in Chap. 3 we argued that RE may not be available in sufficient annual quantity to replace existing levels of FF use, at least in a sustainable manner. Second, in Sect. 4.3, we saw that dynamic energy considerations imply that the rate of introduction of low EROI sources, with high upfront energy input costs, both of which characterise most RE sources, is subject to constraints. The world does need to replace FFs with RE sources, but the decades likely needed for this replacement rule out RE as having more than a minor solution to play in overcoming our immediate environmental crises [34]. Energy conservation then appears as the only solution for rapid GHG reductions left in the short term.

The fourth theme concerns global inequality, whether it is in per capita incomes, energy use, GHG emissions, medical services, education, and the like. As an example of inequality, Auerback [3] pointed out that the burden of the Covid-19 pandemic in the US is disproportionately borne by minorities and the poor. In low-income countries, the pandemic has forced many more households into poverty. At the international level, income inequalities are often shrinking, but are still usually rising at the *intra-national* level [33]. Another example, The UN High Commissioner for Refugees (UNHCR), estimates that one out of every 100 persons are now displaced from their homes, a 2-fold rise since 2010 [51]. These inequalities are an important contributing factor to the fragile international and national political systems that exist today. Unfortunately, as several researchers have pointed out, many proponents advocating reductions in GHG emissions feel that greater equality is a barrier to effective climate action. On the other hand, Klinsky et al. [21] titled their paper 'Why equity is fundamental in climate change policy research.' The Sustainable Development Goals (SDGs) of the UN [50] do stress equity in their list of goals. The authors of the highly influential 'global limits' concept now stress the need for a just as well as ecologically sustainable trajectory for Earth [14].

However, as Jason Hickel [16] has stressed, many of the UN SDGs are in conflict with each other. These conflicts can be traced to a profound problem pointed out by many other authors as well, for example [10, 42]: continued economic growth on a finite planet will undermine biosphere integrity. One view is that we ought to target ecological limits rather than economic growth. The point is well-taken, but it is likely that any attempt to seriously avoid transgressing any of the ecological limits will result in global GDP decline—at least as GDP is currently understood and measured. The pandemic has also meant that the date set for achieving the SGD goals has been pushed back even further into the future.

One way of viewing the challenges we face is shown schematically in Fig. 5.1. The human consumption curve, which is a measure of the resources necessary for all humans to live a full and productive life, can, for simplicity, be assumed constant per capita. (Actual global human consumption is, of course, higher.) As the UN expect global population to continue its rise, it is displayed as rising over time. The upper ecosystem services curve shows the ability of the biophysical resources—the Earth's natural capital—to provide ecosystem services. Our present consumption of 1.7 Earths is already thought to have pushed us into an ecological overshoot, whereby the decline in the Earth's natural capital is such that our continued consumption levels could soon lead to its decline being unrecoverable, and so diminishing what is available for us to consume (Fig. 5.1a). Avoiding this will require us to significantly reduce our consumption to the level at which the Earth's natural capital can recover-although such a recovery will be difficult. Sustaining ecosystem services with population growth will require continued reductions in global per capita consumption. While it is possible that with technological advances a fixed biophysical capacity could satisfy the needs of an ever-rising number of people, the arguments presented in Chap. 2 cast doubt on this claim. Besides any technological advances, we argue that not only will significant changes in global consumption levels be required but that these changes must also be accompanied by greater equity, given the present inequitable distribution in consumption. Many still do not have access to such basic needs as an adequate diet.

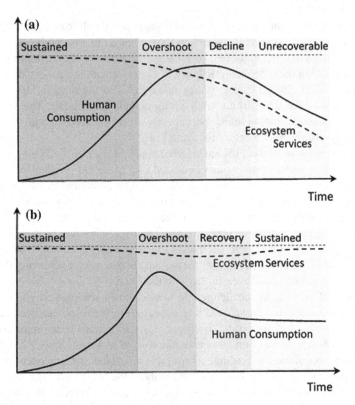

Fig. 5.1 Schematic diagram showing shrinking space between Earth's biophysical capacity and human needs. **a** Human consumption overshoots biophysical capacity followed by an unrecoverable decline in natural capital represented by ecosystem services. **b** Overshoot followed by a reduction in human consumption sufficient to recover natural capital; ecosystems services recover. *Sources* [32, 42, 48]

5.2 Changes Needed for Equity and Ecological Sustainability

Chapters 2 and 3 together examined all the technological solutions promoted for solving the climate emergency, and even for mitigating the other environmental damages caused by energy use. For various reasons, they were all found wanting. But what if many, if not all, were used in concert? Such a solution seems reasonable, since it raises the possibility of each approach being used at a lower level, below a possible cut-off point for serious adverse environmental (or social) effects. This combined approach is already implemented to some extent, and in any case, different countries have different opportunities for each technology. But, as we have shown for the urban context [31], the various approaches can also conflict with each other, limiting their combined use. For instance, as shown in Sect. 3.4.1, SRM will lower the effectiveness of some passive solar energy measures, as well as lower output from solar thermal electricity plants.

As we have argued, the world has to use less energy. Makiereva et al. [24] have earlier arrived at a similar conclusion from a very different—and controversial—starting point. Their conclusion: 'There is no logic in substituting fossil fuels by some pollution-free renewable energy sources. It is necessary to reduce the very magnitude of energy consumption and the related magnitude of the anthropogenic pressure on the biosphere.' Energy efficiency proponents also promote energy reductions, but we are suggesting something very different. Energy cuts through efficiency improvement seek to reduce the energy use of each energy-using device, but not the number of devices, or even their intensity of use. But as shown in Sect. 2.3, any energy efficiency gains for devices have feedback effects which swamp any savings, especially by increasing ownership in industrialising countries, which implicitly use OECD countries as a template for their future consumption. Also, road transport has heavy ecological costs quite independent of its energy use (see Sect. 2.3). The remainder of this section looks at possible approaches to achieve rapid and deep energy cuts, by examining three areas (FF subsidies, passenger transport, and food and agriculture), then puts these specific changes in context by discussing the social and economic changes underpinning them.

5.2.1 Remove All Subsidies from Fossil Fuels

Based on known proven reserves and current commodity prices, the total value of the world's fossil fuel reserves is estimated to be worth more than USD 100 trillion, with oil taking the smallest share [6]. Much of this net worth is based on the receipt of substantial subsidies. Some of these can take the form of government subsidies to producers and/or consumers. But the largest subsidy comes in the form of unpaid external costs; these externalities are dominated by the unpaid costs of carbon emissions. Another important (but indirect) component is the costs of road traffic casualties, air pollution, and costs of roads and parking infrastructure, net of road taxes. Coady et al. [7] assessed the total global FF subsidy in 2015 at US$5.3 trillion, or 6.5% of global GDP. Even this figure omitted a possible inclusion: much of the vast military expenditures of the Western countries in the oil-rich Middle East could be seen as necessary for guaranteeing continued oil imports.

Direct subsidies to FF industries should be removed as soon as possible. Removal of the carbon emission subsidy would require much higher prices, perhaps through a carbon tax. According to Wagner et al. [55], a carbon tax 'puts a monetary value on the harms of climate change, by tallying all future damages incurred globally from the emission of one tonne of carbon dioxide now'. Such taxes are already levied on FFs in a number of countries, but the levels are low, and so far, have had little impact on FF consumption, as shown by BP's FF energy statistics [6]. It is vital that the implementation of any carbon tax does not further exacerbate energy inequality, as FF subsidies are often important for low-income households. Some proposals would even improve equity by redistributing a portion of the carbon taxes either nationally or globally [33]. An example is Canada's carbon tax, presently set at only $30 per tonne of carbon, but with an increase to $170 planned by 2030. Those with incomes at the lowest two-thirds receive a quarterly 'carbon dividend', which ensures they get back more than they paid in carbon taxes [56].

The level at which carbon taxes should be set is related to the much-discussed concept of the *social cost of carbon* (SCC), which aims to assess the climate change costs of FF combustion. Various values for SCC have been calculated by economists, but they show an enormous range. In 2013, an official US working group fixed on a range $15–75, but the Trump administration dropped it to $1–7 from 2017. In contrast, Germany's 2020 guidance paper presented two values: $235 and $820 [55]. In light of the risk of CCC discussed in Chap. 1, even higher values have been suggested, all the way up to infinity. For various reasons, some researchers believe that carbon markets should not be part of the solution. Ball [4] has argued that carbon pricing, once perceived by policymakers as a panacea, is becoming a narcotic; it gives the false impression that something is finally being done to halt CC, while sapping the will for alternative measures.

5.2.2 Set Limits on Vehicular Passenger Transport

We will illustrate our approach using passenger transport as an example, as one criticism of global decarbonisation scenarios is the lack of detail on strategies needed, including in the transport sector [22]. Global transport, both passenger and freight, is important for both energy use and CO_2 emissions, with 2018 energy use 121.0 EJ (29.1% of final energy demand) and around 24% of energy-related CO_2 [27]. Our approach first asks the question: What is transport for? [36]. Transport planners often assume that their task is to provide vehicular *mobility*, and at the highest speeds consistent with passenger comfort and safety. Passenger mobility is measured in passenger-km (p-km): ten travellers each making a 5 km trip generate a total of 50 p-km, and so on. Since the advent of motorised public transport, then car travel, and later air travel, the world has witnessed a startling growth in travel, with a growth factor since the year 1900 the authors have estimated as 225 [30]. Figure 5.2 shows the growth of global travel by mode since 1900, with the especially rapid growth seen since the mid-twentieth century mainly due to car and air travel.

But is mobility really what people are looking for when they take trips? Or, is what they really want from travel *access*: access to workplaces, shopping centres, friends, entertainment centres, and so on? Access will certainly involve mobility, but can a given level of access be provided at much lower levels of mobility, and consequently lower transport energy use? One difficulty is that vehicular p-k is easy to measure, whereas access is not, as it depends to some extent on the individual. Vehicular travel levels per capita are so high in OECD countries because we have ignored many of the costs of motorised travel, especially by car. Think of the global road casualty toll, the GHG emissions, community severance, and noise and air pollution, especially

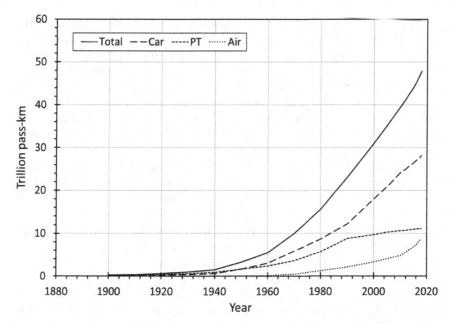

Fig. 5.2 Global vehicular travel by mode, 1900–2018. PT is travel by public transit modes. *Sources* [27, 33, author estimates]

in cities. Sovacool and colleagues [47] have recently surveyed some of these costs for various transport modes, finding that the mean value of estimates for all road passenger transport was (in 2018 values) USD 0.24 per p-km, although estimates varied greatly, from USD 0.06 to USD 2.39 per p-km.

Because of these costs, the first author has tentatively advocated a nominal 4000 p-km per capita limit to *vehicular* travel, in line with similar proposals to limit oil use and energy overall [27]. Non-motorised travel would not be restricted—indeed its use would be encouraged; even now, it is already encouraged for health reasons. One approach for achieving this aim would be through a reversal in the usual ranking of passenger transport modes, with non-motorised modes (walking and cycling) receiving priority, then public transport, and finally private travel, as argued in our earlier work [26, 35]. There is a strong equity component in this idea, since in presently low-income countries, most road casualties are pedestrians and cyclists, not vehicle occupants [28]. Priority for these non-motorised modes, and restricting vehicular road travel, would favour such low-income travellers.

How could such policies be implemented, given the entrenched position of private travel among both the public and their elected representatives? One answer is that some countries and cities are already taking steps to limit car travel, particularly in urban centres [30]. Such changes can be justified by improvement to the quality of urban life, lower noise and air pollution, fewer road casualties, and the increased opportunity for physical exercise.

What has happened is that we have increased the *convenience* of car travel relative to other modes, and largely ignored its far higher environmental and human costs. Urban residents thought that by switching from public to private travel, they would save door-to-door travel time for a given trip. These time savings usually occurred—at least for a while—unless traffic congestion heavily impacted car speeds. But with mass car travel, preferred destinations moved further from home, as for example, walking trips to local shops were replaced by vehicular trips to more distant drive-in shopping centres. In addition, the cost structure of private travel encouraged reliance on the car for all trips, because of low perceived operating costs.

In OECD countries, progressive de-industrialisation has meant a decline in factory jobs and a corresponding increase in service employment. The rise of more jobs in the service economy, such as in health, education, and social services, means that workplaces are far more evenly distributed across the city than were factory jobs, as these service workplaces tend to be located near the populations they cater to for. Urban travel per capita, *ceteris paribus*, should have declined. Paradoxically, with the parallel transition from public to private travel, per capita travel instead rose several-fold, at least in Australian cities [26]. In the high-income OECD countries, we have entered the era of *hypermobility* [30]. If, however, travel convenience was cut in the ways suggested, urban land uses and residents would adjust to the new realities, with more use of local services and workplaces [26].

With the changes in travel convenience suggested here, such as the imposition of parking space restrictions, inner city road closures, or speed limit reductions, much travel reduction could occur quickly, although, in other cases, years might be needed. One other, indirect, method often suggested for cutting urban travel—very high increases in urban density—would cut car travel, but would take decades to implement, and could prove even more unpopular with city residents than more direct measures to lower travel convenience. The changes suggested above do not rely on carbon taxes. Sufficiently high carbon taxes could lower travel (and travel fuel use) to any desired level, but would prove inequitable in many OECD countries, given the high rate of car ownership for lower income groups in countries like Australia, Canada, and the US.

5.2.3 Change Our Diet and Agriculture Practices

Another important change that would provide far-reaching benefits is a shift to a more vegetarian diet, not only in OECD countries but also in countries like China, with a rising share of animal products in their diet. Globally, food-related GHG emissions are large, as agriculture not only produces CO_2 emissions but also CH_4, especially from ruminant animals, and N_2O emissions from soils. If animal products consumption continues to rise, emissions are forecast to grow from 7.5 Gt CO_2-e (CO_2 equivalent)—about 20% of all present global emissions—to 11.35 Gt CO_2-eq in 2050. For red meat and poultry alone, the equivalent figures were 4.6 and 7.1 Gt CO_2-eq, respectively. As is the case for energy conversion in general, useful energy is lost at each conversion stage. Animal products are at a higher trophic level than plant food, but conversion losses are magnified by the practice of feeding livestock with grain, soybeans, and even fish flour: they thus produce more GHGs per kilojoule of food energy actually consumed by humans. Already, 36.9% of the world's total grain harvest in 2017–2019, estimated at some 2.68 Gt annually, was fed to animals [13]. If fed directly to humans, this grain could provide an additional quantity of around one kg per person per day. If no animal-based food was consumed at all, 8 GtCO$_2$-eq of GHGs could be avoided each year, with lower savings for some meat and dairy use [44].

An added benefit of a shift away from meat is the opportunity it would afford to increase biodiversity by restoring gazing lands, which are estimated to occupy around 33 million km^2, or 25% of all available ice-free land [13]. GHG emissions associated with land clearing would also fall, critically for endangered regions such as the Amazon, which is at significant risk from deforestation for cattle farming [52]. It could even lessen the risk of future pandemics, as both bushmeat consumption and intensive animal raising increase the risk of dangerous viruses crossing over from animals to humans. Finally, as our discussion in Sect. 3.4.3 showed, shifting to a more vegetarian diet could allow more bioenergy to be produced—but only after adequate nutrition is provided for all humans. Particularly for OECD countries with protein intake already more than adequate, reduced red meat consumption could also improve health.

Not only what type of foodstuffs are produced but also the way food is produced is important. Even today, agricultural systems can be usefully classified as pre-industrial, semi-industrial, and full-industrial. The food output per unit of energy input is much higher for the first two systems than for the third, where it is little more than unity [36]. And, just as we saw for non-food energy production in Chap. 3, industrial agriculture produces significant ESME costs, which again are largely ignored. These costs include pollution from fertiliser and pesticide runoff, declining water tables, salinisation, and desertification. Industrial agriculture ranks much higher than pre-industrial agriculture in terms of land efficiency in the form of higher yields (tonnes/hectare), but at the expense of energy efficiency (food output/energy input). Were energy efficiency and CO$_2$ emission limitation considered of vital importance, yields would likely need to be reduced. But then, even more land, including forested land, would be needed for global agriculture. Clearly, there is a need to strike a balance for minimising overall environmental damage.

Agriculture is not only a means to produce food; it is also a way of life for hundreds of millions of people. According to McMichael [25], agriculture serves *multi-functional* purposes; in addition to food production, it is also valued for 'its contribution to ecosystem management, landscape protection, rural employment, fostering farming knowledge, rural life, cuisine maintenance, and regional heritage'. Unfortunately, in the light of the above discussion on the need for vegetarianism, this multi-functionalism also extends to much animal product farming, such as the animal herders in East Africa and other regions, compounding the difficulty of change.

There is yet another way in which a change in food practices could save both GHG emissions and energy use in agriculture, and that is by eliminating food wastes. In low-income countries, food losses occur mostly immediately post harvest. In high-income countries, in contrast, Atkinson [2] has found that households waste 50% of their food. For China, researchers Xue and colleagues [59] have recently found that around 350 Mt, or 27% of annual food production, is either lost or wasted. Some 45% of this loss occurred in the post-harvest phase. They further showed that 'the land, water, carbon, nitrogen, and phosphorus footprints' associated with this food waste are very significant, with the carbon footprint greater than that for all sectors in the UK. Other sources of waste occur in the food industries, and in food services industries such as restaurants [39]. Processing of foods, particularly in OECD countries, is an important source of food waste [59], and processed foods also use more energy per kilojoule delivered to the table than unprocessed food.

5.2.4 Practice Degrowth and Improve Equity

The past decade or so has witnessed the rise of a new term, *degrowth*, although similar ideas have been around for much longer, in the works of researchers such as Herman Daly [e.g. 10]. The degrowth idea centres on the planned reduction in national GDP in high-income economies. A common assumption in the degrowth literature is that the multiple dangers facing us are incapable of being solved in the context of continuing economic growth. Specifically, degrowth writers are sceptical that economic growth can be decoupled from energy use—see also Parrique et al. [40].

Figure 5.3 demonstrates how tightly global commercial primary energy use and global GDP have been for the past four decades, suggesting that breaking the link will be difficult. It is easy to underestimate the magnitude of the changes required. Average commercial energy consumption per capita in 2020 varied from around 600 GJ in several high-income countries to about 4.5 GJ in East and Central Africa [6]. The world average was 71.4 GJ/capita, which produced about 32 Gt of CO_2, roughly distributed between countries by their energy use. For sustainability, it is clearly not possible for the rest of the world to match even present rich country per capita levels of energy use, CO_2 emissions, car ownership, air travel, and so on. Yet equity demands a more equal distribution of the Earth's resources. The only way forward is to abandon economic growth as an objective. As Daly [10] has argued: 'Growthism is consuming the life support capacity of the biosphere for the benefit of a small minority of the present generation, while shifting the real but uncounted costs on to the poor, future generations, and other species.' As illustrated in Fig. 5.1, failure to rein in consumption may lead ultimately to an unrecoverable decline in the Earth's biocapacity, followed by an ever-diminishing ability to appropriate the shrinking natural capital.

Degrowth need not mean a decline in OECD living standards, in the sense of lower satisfaction of human needs [42]. The GDP was not designed as a measure of human welfare, but as an indicator of the level of economic activity in a country. Although—at least up to the year 2019—GDP per capita continued to grow in OECD countries,

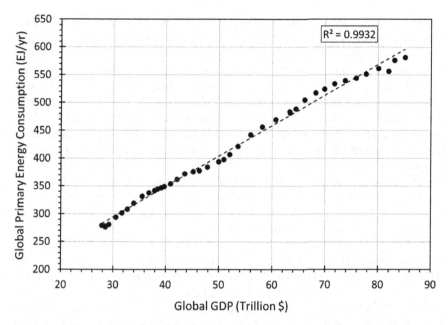

Fig. 5.3 Global annual primary energy consumption versus global GDP (constant USD 2010 values) 1970–2020. Correlation coefficient shown for line of best fit. *Sources* [6, 58]

another index, more attuned to human needs and to equity, the genuine Progress Indicator, has steadily fallen since the mid-1970s [17]. Further, market economies are poor at providing goods and services that are not valued by the marketplace, resulting in the under-provision of these goods and services. Many of these are important for the quality of life. Studying lifestyles and practices of a century ago in now-wealthy economies could provide clues for the way forward. The high-income societies would also do well to study the more creative measures by which low-income, low-energy societies of today manage with low consumption levels.

A common argument is that global economic growth is needed to improve the life chances for presently low-income societies. At present, as discussed in Sect. 1.3, most global economic growth is further enriching those already well-off. The changes already discussed in this section will help to improve equity. If high levels of carbon taxes are adopted, a share should be used to support low-income societies. A universal basic income should also be at least considered as a safety net [33].

5.3 Are the Required Changes Feasible?

How will key decision-makers respond to the multiple biophysical and socio-political crises our world faces? Are the suggested solutions socially and politically feasible? Will these solutions be sufficient? Feasibility may be the greatest uncertainty of all. If we simply extrapolate from trends of the past few decades, the future looks very grim indeed. Carbon emissions from energy use and deforestation would continue their rise annually, other biophysical limits will run an increased risk of being breached, and global inequality will climb to even higher extremes. It does not have to be this way. The historical record shows that momentous social and political changes for the better have occurred, even in the past few decades. Think of our changing attitudes towards race, sexual preference, the mentally ill, or treatment of animals. Seibert and Rees [45] likewise provide a note of optimism regarding tackling climate change, by cautioning that 'history is replete with stellar achievements that have arisen only from a dogged pursuit of the seemingly impossible.' In this section, we first look at the several factors hindering strong action on mitigating CC. Nevertheless, some observers have seen several encouraging signs for the feasibility of strong social/economic change. We examine these various possible drivers, discussing both the strengths and weaknesses of their arguments in the following paragraphs.

At first glance, CC mitigation should have been easy. A survey in 2019 of climate scientists showed that a full 100% considered climate change to be real, up from 97% a few years early [41]. What about the general public? In Australia, a nationally representative survey conducted in April 2021 found that 60% believe that 'global warming is a serious and pressing problem' and that we should act now. An even larger share (80%) want the government to adopt a zero-emissions target by 2050 [23]. This, despite the fact that Australia is a large user and exporter of both coal and natural gas. An earlier poll, soon after the May 2019 federal elections, also found that a large majority supported action on climate change, with the support strongest among the more educated, those with progressive politics, and younger people [8].

A survey published by the European Commission in July 2021 [12] found that Europeans consider 'climate change is the single most serious problem facing the world', with 93% regarding it as a serious problem, and 78% considering it 'very serious'. In the same survey, respondents ranked climate change as the most serious global problem (18%), followed by 'deterioration of nature' (7%), and health problems due to pollution (4%). Thus, 29% overall believed that one of the three limits identified in Sect. 1.2 was the most important challenge the world faces. These figures help explain why, in Europe, political parties across the political spectrum support climate change action. Greta Thunberg from Sweden, with her school strikes for climate action, has inspired many young people throughout the world.

In the US, the position is more complex. Only 50% of voters in an April 2021 survey believed climate change was a critical threat to US interests, but this figure is still up 10 percentage points on a 2017 poll. A further 27% thought climate change was important but not critical. (However, the use of the term 'US interests' may precondition those surveyed to respond in a *realpolitik* way. The national interest term was not used in the Australian or EU polls.) Voters are also more polarised by party preference than elsewhere: Three-quarters of Democrats judged CC as a 'critical threat', an increase of 16 points over a 2017 poll, compared with only around one-fifth of Republican voters (with only the rise of one point over 2017) [19]. A change of government could once again see the US leaving the 2015 Paris climate agreement.

A 2015 survey of Facebook users in 31 countries found large variations between countries, but generally higher rates of CC belief in OECD countries [57]. Governments worldwide, not only in the OECD, at least nominally support action on CC, notably with the international Paris Agreement signed in 2015 by 196 parties to limit GHG emissions. The European Union binds its member nations to even more stringent emission cuts. The US, under President Biden, has re-entered the Paris Agreement, and has pledged strong climate action, as has China, the leading CO_2 emitter. Individual countries, such as Sweden and the UK's conservative government, have stated aims which promise even more ambitious targets [1].

Given this level of support, what holds us back from effectively tackling climate change? According to Anderson and colleagues [1], when for international equity reasons, the 'differentiated responsibilities' of OECD countries are considered in line with the Paris Agreement, the reality is very different from the promises. The allowable carbon budgets of Sweden, the UK, and other developed countries would be halved. All would need to increase their rate of emissions reduction to an unprecedented 10% per annum, if the 1.5 °C target is to be met, unless unproven schemes such as BECCS are adopted on a very large scale. The big challenge concerns exactly how to divide the ever-shrinking carbon pie, as discussed in Chap. 1. The key difficulty is that, historically, industrial countries account for most CO_2 emissions, but today the industrialising countries are responsible for around two-thirds; their emissions, no matter how modest on a per capita basis, can no longer be ignored. Arguments at the international level about national emission reduction responsibilities act to prevent rapid agreement. The end result is that the world is not on target to avoid catastrophic CC.

Several reasons could help explain this inertia. One, as already mentioned, is the 'free-loader' problem. If one country makes an extraordinary effort to cut emissions, all countries will benefit equally—albeit to a very small extent. So, unlike local pollution, individual governments find it difficult to get support for serious climate mitigation, with the perceived sacrifices they entail. Also, despite the views of climate scientists, and most of the public, a substantial minority of people in OECD countries still do not think CC is a serious problem. The gradual change of around 0.1 °C per decade is hard to perceive, partly because of natural climate variability, let alone daily weather variations in temperate climate regions. For these climate sceptics, the signal-to-noise ratio is not enough. FF energy organisations and exporting countries, and even energy-intensive industries, obviously have a vested interest in the continuation of the present energy FF-dominated system [49], and have at times supported climate sceptic organisations. For them, climate change is, in Al Gore's words, 'an inconvenient truth'.

Furthermore, as Chaps. 2 and 3 discussed in detail, many technical fixes have been proposed for mitigating climate change, and will continue to be proposed. Whether feasible or not, the plethora of options does blunt the urgency for change now. Witness the popularity of 'overshoot scenarios', where action can be deferred for the future—and made Someone Else's Problem! For this reason, CDR and SRM proposals are also often criticised for sapping the will for more direct climate action such as deep cuts to FF use.

Amid this inertia on CC action, some have still seen encouraging signs. One, paradoxically, is the recent rise in severity and frequency of extreme weather events around the world, as documented in Sect. 1.4. It might be expected that this 'climate forcing' would lead countries to enact stricter climate mitigation measures. Some evidence for this view comes from the US. One study found that countries that had directly experienced extreme climate events were more likely to support action on CC, but as found earlier, support was mediated by political party affiliation [9].

Others have seen the ongoing Covid-19 pandemic, and the ensuing social and economic changes, as an opportunity for re-thinking economic and social fundamentals [e.g. 5]. In 2020, the global economic downturn did lead to a 4% decline in global energy use [18]. In the first half of 2020, as the first wave of pandemic infections impacted the world, some commentators were optimistic that the rapid social changes such as wearing face masks, practicing social distancing, and remote learning and conferencing during lockdowns signalled the beginning of widespread social change. Others pointed out the similarities in the pandemic and global climate change, and, presumably, the approaches needed to tame them both. Both result from human actions, and their impacts are global, with similar societal impacts. In both cases also, it is low-income households that are the worst affected [20]. But, as shown in Chap. 1, early hopes for immediate change to CC mitigation using the rapid pandemic response as a template have not been realised. Nevertheless, the pandemic has led to a growing realisation that all of Earth's people are in this together, facing a common danger. Unvaccinated populations will lead to fresh outbreaks from new strains, so that it is in the self-interest of rich countries to ensure vaccination in low-income countries [11]. The alternative is to shut down movement between these countries, but this could ultimately come at a considerable economic cost, as many rich countries rely on these low-income countries as a source of labour.

There is another lesson being learned the hard way from the pandemic, which has enormous relevance for CC mitigation: it is better to act immediately with radical measures such as lockdowns to control outbreaks than to dither around with politically easier but ineffective measures. This is because, if unchecked, the infected cases rise exponentially. Similarly, as shown in Chap. 1, the harmful effects of CC are non-linear with temperature rise. The damages from extreme weather (bushfires, heatwaves, floods) are already beginning to affect us all, not only groups like tropical country residents.

Otto et al. [38] are optimistic about the possibilities for change in climate. They have identified and examined six of what they term 'social tipping interventions', which are a response to the risks from tipping points in the climate system discussed in Chap. 1. The six items were identified both from a review of the relevant literature and views of experts. They include actions such as removing fossil fuel subsidies, as above, and building carbon–neutral cities, as well as giving information not only on emissions from specific products and services but also on the moral implications of FF use. Whether these interventions of this type could produce the profound changes needed remains to be seen, and we must add to these the need to continually adapt our behaviour and reset our expectations to the ever-evolving challenges that lie ahead.

5.4 Summary and Conclusions: Switching Off

By late 2021, all signs indicated that economies were still fixated on economic growth and were prepared to tolerate the often-hidden environmental risks of such a path. As shown in Chap. 1, global energy consumption appears set to continue its pre-pandemic growth path [18]. On the one hand, there is the need to drastically reduce energy, especially from FFs, for ecological sustainability reasons, and on the other, the need for some energy growth in low-energy societies.

The world is being pulled in several directions. The dominant direction is for continued economic growth at (almost) all costs. True, this is usually only implicit, but the statistics bear it out: global GDP continues to rise, barring downturns caused most recently by the 2008 Global Financial Crisis and the ongoing Covid-19 pandemic. Meanwhile, ecological indicators on the health of the planet continue to worsen. The argument is that economic growth is needed for the world's poor, although as mentioned in Chap. 1, the lion's share of the growth is going to the already wealthy.

The second direction is for 'Green Growth' or the 'Green New Deal' as it is termed in America [45]. In this approach, economic growth need not conflict with ecological stability. For its proponents, green growth can give a new impetus for economic growth, based on novel green technologies, particularly renewable energy. Chapter 3 showed the difficulties with this view when a comprehensive Earth Systems Science approach is adopted that allows for full consideration of all the impacts of energy systems. Renewable energy is not as green as is made out, and when it is made fully green, the quantity of energy globally available is probably much less than current—let alone projected future—global energy levels.

Climate scientist, Corrine Le Quéré [53], interviewed in 2020 in *New Scientist*, urges that even if–as looks increasingly likely, as even the IPCC now acknowledges—the world misses the 1.5 °C target, we should then work hard to stay below the 2 °C target. Even if we go beyond 2 °C, there is no point in throwing in the towel: the world at 2.5 °C will be a much more liveable place than a world at 3 °C, and so on. Or, as the latest IPCC report stresses, there is still time to act to avoid the worst climate outcomes. At no point do we give up. But attempts to mitigate CC (or other pressing ecological problems) must not be made at the expense of the less well-off, which, as Sovacool [46] has argued, is the present case. Our greatest challenge will be to determine what we must switch from doing, if we ourselves are not to be the ones who are switched off.

References

1. Anderson K, Broderick JF, Stoddard I (2020) A factor of two: how the mitigation plans of 'climate progressive' nations fall far short of Paris compliant pathways. Clim Pol 20(10):1290–1304. https://doi.org/10.1080/14693062.2020.1728209
2. Atkinson A (2014) Urbanisation: a brief episode in history. City 18(6):609–632

3. Auerback M (2020) Why COVID-19 is the great unequalizer: the pandemic's impact is being experienced disproportionately by minorities and the poor. Real-World Econ Rev 92:252
4. Ball J (2018) Hot air won't fly: the new climate consensus that carbon pricing isn't cutting it. Joule 2:2491–2494
5. Benach J (2021) We must take advantage of this pandemic to make a radical social change: the coronavirus as a global health, inequality, and eco-social problem. Int J Health Services 51(1):50–54. https://doi.org/10.1177/0020731420946594
6. BP (2021) BP statistical review of world energy 2021, 70th edn. BP, London
7. Coady D, Parry I, Sears S et al (2017) How large are global energy subsidies? World Dev 91:11–27
8. Colvin RM, Jotzo F (2021) Australian voters' attitudes to climate action and their social-political determinants. PLoS ONE 16 (3): e0248268 https://doi.org/10.1371/journal.pone.0248268
9. Cutler MJ, Marlon J, Howe P et al (2020) 'Is global warming affecting the weather?' Evidence for increased attribution beliefs among coastal versus inland US residents. Environ Sociol 6(1):6–18. https://doi.org/10.1080/23251042.2019.1690725
10. Daly H (2019) Growthism: its ecological, economic and ethical limits. Real-World Econ Rev 87:9–22
11. Dhai A (2021) Access to COVID-19 vaccines as a global public good: a co-ordinated global response based on equality, justice and solidarity is key. S Afr J Bioethics Law 314(1):2–3
12. European Commission (2021) Eurobarometer Survey: Europeans consider climate change to be the most serious problem facing the world https://ec.europa.eu/commission/presscorner/detail/en/ip_21_3156
13. Food and Agriculture Organization (FAO) (2020) World food and agriculture - statistical yearbook 2020 https://doi.org/10.4060/cb1329en
14. Folke C, Polasky S, Rockström J, et al (2021) Our future in the Anthropocene biosphere. Ambio 50: 834–869 https://doi.org/10.1007/s13280-021-01544-8
15. Godfray HCJ, Aveyard P, Garnett T, et al (2018) Meat consumption, health, and the environment. Science 361: (6399): eaam5324 https://doi.org/10.1126/science.aam5324
16. Hickel J (2019) The contradiction of the sustainable development goals: Growth versus ecology on a finite planet. Sustain Dev 27:1–12
17. Hickel J, Brockway P, Kallis G, et al (2021) Urgent need for post-growth climate mitigation scenarios. Nature Energy https://doi.org/10.1038/s41560-021-00884-9
18. International Energy Agency (IEA) (2021) Global energy review 2021. https://www.iea.org/reports/global-energy-review-2021
19. Jenkins LM (2021) Half of U.S. voters now characterize climate change as a 'critical threat' https://morningconsult.com/2021/04/27/paris-agreement-climate-change-threat-poll/
20. Joshi M, Caceres J, Ko S et al (2021) Unprecedented: the toxic synergism of Covid-19 and climate change. Curr Opin Pulm Med 27:66–72
21. Klinsky S, Roberts T, Huq S et al (2017) Why equity is fundamental in climate change policy research. Glob Environ Change 44:170–173
22. Loftus PJ, Cohen AM, Long JCS et al (2015) A critical review of global decarbonization scenarios: What do they tell us about feasibility? WIREs Clim Change 6:93–112
23. Lowy Institute (2021) Climate Poll-2021 https://www.lowyinstitute.org/publications/climatepoll-2021
24. Makarieva AM, Gorshkov VG, Li B-L (2008) Energy budget of the biosphere and civilization: Rethinking environmental security of global renewable and non-renewable resources. Ecol Complex 5:281–288
25. McMichael P (2011) Food system sustainability: questions of environmental governance in the new world (dis)order. Glob Environ Change 21:804–812
26. Moriarty P (2016) Reducing levels of urban passenger travel. Int J Sustain Transp 10(8):712–719. https://doi.org/10.1080/15568318.2015.1136364
27. Moriarty P (2021) Global transport energy. Encyclopedia 1: 189–197 https://doi.org/10.3390/encyclopedia1010018

28. Moriarty P, Honnery D (2008) Low-mobility: the future of transport. Futures 40:865–872
29. Moriarty P, Honnery D (2014) The Earth we are creating. AIMS Energy 2(2):158–171
30. Moriarty P, Honnery D (2019) Prospects for hydrogen as a transport fuel. Int J Hydrog Energy 44:16029–16037
31. Moriarty P, Honnery D (2019) Creating environmentally sustainable cities: not an easy task. In: Archer K, Bezdecny K (eds) Handbook of emerging 21st century cities. Edward Elgar, London
32. Moriarty P, Honnery D (2019) Ecosystem maintenance energy and the need for a green EROI. Energy Pol 131:229–234
33. Moriarty P, Honnery D (2020) New approaches for ecological and social sustainability in a post-pandemic world. World 1:191–204. https://doi.org/10.3390/world1030014
34. Moriarty P, Honnery D (2021) The limits of renewable energy. AIMS Energy 9(4):812–829
35. Moriarty P, Honnery D (2021) Reducing personal mobility for climate change mitigation. In: Lackner M, Sajjadi B, Chen WY (eds) Handbook of climate change mitigation and adaptation (3/e). Springer, NY
36. Moriarty P, Honnery D (2021) Reducing energy in transport, building and agriculture through social efficiency. In: Lackner M, Sajjadi B, Chen WY (eds) Handbook of climate change mitigation and adaptation (3/e). Springer, NY
37. OECD/Food and Agriculture Organization (FAO) (2020) OECD-FAO agricultural outlook 2020–2029 OECD, Paris. https://www.oecd.org/publications/oecd-fao-agricultural-outlook-19991142.htm
38. Otto IM, Donges JF, Cremades R et al (2020) Social tipping dynamics for stabilizing Earth's climate by 2050. PNAS 117(5):2354–2365
39. Parfitt J, Barthel M, Macnaughton S (2010) Food waste within food supply chains: quantification and potential for change to 2050. Phil Trans R Soc B 365:3065–3081
40. Parrique T, Barth J, Briens F, et al (2019) De-coupling Debunked. Eur Environ Bur eeb.org/decouplingdebunked
41. Powell J (2019) Scientists reach 100% consensus on anthropogenic global warming. Bull Sci Technol Soc 37(4):183–184
42. Rees WE (2021) Growth through contraction: conceiving an eco-economy. Real-World Economics Rev 96 http://www.paecon.net/PAEReview/issue96/Rees96.pd
43. Ross AGP, Crowe SM, Tyndall MW (2015) Planning for the next global pandemic. Int J Infect Dis 38:89–94
44. Schiermeier Q (2019) Eat less meat: UN climate change panel tackles diets. Nature 572:291–292
45. Seibert MK, Rees WE (2021) Through the eye of a needle: An eco-heterodox perspective on the renewable energy transition. Energies 14: 4508 https://doi.org/10.3390/en14154508
46. Sovacool BK (2021) Who are the victims of low-carbon transitions? Towards a political ecology of climate change mitigation. Energy Res Soc Sci 73: 101916
47. Sovacool BK, Kim J, Yang M (2021) The hidden costs of energy and mobility: a global meta-analysis and research synthesis of electricity and transport externalities. Energy Res Soc Sci 72: 101885
48. Spangenberg J H (2014) Institutional change for strong sustainable consumption: sustainable consumption and the degrowth economy. Sustain: Sci, Practice, Pol 10 (1): 62–77
49. Treen KMd, Williams HTP, O'Neill SJ (2020) Online misinformation about climate change. WIREs Clim Change 11: e665 https://doi.org/10.1002/wcc.665
50. United Nations (UN) (2020) The sustainable development goals report. 2020. https://unstats.un.org/sdgs/report/2020/The-Sustainable-Development-Goals-Report-2020.pdf
51. United Nations High Commissioner for Refugees (UNHCR) (2019) Global trends: forced displacement 2019 https://www.unhcr.org/5ee200e37.pdf
52. Vaughan A (2020) Fifth of Brazilian beef exports to EU linked to illegal deforestation. 2020 https://www.newscientist.com/article/2249083-fifth-of-brazilian-beef-exports-to-eu-linked-to-illegaldeforestation/
53. Vaughan A, Le Quéré C (2020) The climate fight after coronavirus. New Sci 247:36–39
54. Wackernagel M (2018) Day of reckoning. New Sci August 4: 20–21

55. Wagner G, Anthoff D, Cropper M et al (2021) Eight priorities for calculating the social cost of carbon. Nature 590:548–550
56. Webb R (2021) The climate financier. New Sci 20 March: 44–49
57. Wikipedia (2021) Climate change denial https://en.wikipedia.org/wiki/Climate_change_denial
58. World Bank (2021) GDP (constant 2010 $US) https://data.worldbank.org/indicator/NY.GDP.MKTP.KD
59. Xue L, Liu X, Lu S et al (2021) China's food loss and waste embodies increasing environmental impacts. Nature Food 2:519–528

Printed in the United States
by Baker & Taylor Publisher Services